The Art of Digital Orchestration

The Art of Digital Orchestration explores how to replicate traditional orchestration techniques using computer technology, with a focus on respecting the music and understanding when using real performers is still the best choice.

Using real-world examples including industry-leading software and actual sounds and scores from films, VR/AR, and games, this book takes readers through the entire orchestration process, from composition to instruments, performance tools, MIDI, mixing, and arranging. It sheds light on the technology and musical instrument foundation required to create realistic orchestrations, drawing on decades of experience working with virtual instruments and MIDI.

Bringing together the old and new, *The Art of Digital Orchestration* is an excellent resource for anyone using software to write or compose music.

The book includes access to online videos featuring orchestration techniques, MIDI features, and instrument demonstrations.

Sam McGuire is an active audio engineer, composer, and faculty member of the University of Colorado, Denver. He is the author of multiple audio technology book and numerous online training modules. Sam's research into 360° audio and impulse response technology has taken him to Europe and Asia, recording balloon pops in castles, cathedrals, and concert halls.

Zbyněk Matějů is a contemporary classical music composer. He is well known for his ballet, scenic, and film music, and for his symphony works. He has received many international awards for his compositions.

The Art of Digital Orchestration

Sam McGuire and Zbyněk Matějů

Routledge
Taylor & Francis Group

LONDON AND NEW YORK

First published 2021
by Routledge
2 Park Square, Milton Park, Abingdon, Oxon OX14 4RN

and by Routledge
52 Vanderbilt Avenue, New York, NY 10017

Routledge is an imprint of the Taylor & Francis Group, an informa business

Library of Congress Cataloging-in-Publication Data
Names: McGuire, Sam, author. | Matějů, Zbyněk, 1958- author.
Title: The art of digital orchestration / Sam McGuire and Zbyněk Matějů.
Description: New York : Routledge, 2020. | Includes bibliographical references and index.
Identifiers: LCCN 2020031229 (print) | LCCN 2020031230 (ebook)
Subjects: LCSH: Instrumentation and orchestration. | Computer composition (Music) | Sound recordings--Production and direction. | Software samplers.
Classification: LCC MT70 M35 2020 (print) | LCC MT70 (ebook) | DDC 781.3/4--dc23
LC record available at https://lccn.loc.gov/2020031229
LC ebook record available at https://lccn.loc.gov/2020031230

British Library Cataloguing-in-Publication Data
A catalogue record for this book is available from the British Library

ISBN: 978-0-367-36275-1 (hbk)
ISBN: 978-0-367-36274-4 (pbk)
ISBN: 978-0-429-34501-2 (ebk)

Typeset in Univers
by MPS Limited, Dehradun

Visit the eResources: www.routledge.com/9780367362744

To my beautiful wife who has a beautiful smile, amazing hair, and has made my life better in every way.

Sam McGuire

Contents

Contents

Contents

Contents

Acknowledgements

Thanks to Bob Damrauer and the Office of Research Services at CU Denver for their support in traveling to the Czech Republic for research included in this book. Thanks to the College of Arts & Media at CU Denver for their endless support over the years. This book wouldn't have been possible without their faculty development grant program and counsel.

A huge thanks to Nathan Van der Rest for his writing in Chapters 3 and 6. We really need to plan a synth vacation! Yep, that's a thing. Thanks to all of the other professionals who helped with contributions to this book. It wouldn't be the same without you!

Sam McGuire

Chapter 1

The process of composing

Music technology continues to evolve at breakneck speeds, enabling composers to create more realistic digital orchestrations with less effort. It also enables composers to easily create complex orchestrations which would otherwise take a lot of effort. This book is an exploration of everything related to the creation of digital orchestrations, from MIDI sequencing to mixing the final parts in an audio workstation.

It needs to be said at the very beginning of this journey that there is nothing in here about harmony, compositional forms, or anything related to the musical portion of orchestration except as it applies to the utilization of tools used in the process. The primary objective of every chapter is to provide knowledge about the tools related to creating a realistic digital orchestration. The few exceptions to this include a section about acoustic instruments because they are then compared to their digital counterparts and explanations of things like tempo and the role of arrangements in the mixing process.

The role of the composer

The compositional process is as varied as the composers who engage in it. A hundred composers could describe their individual process and all of them could be different. In spite of this spread, there is a single set of tools and typical work flows. It's important to discuss the technology in relation to the music though, and so a composer simply had to be involved with the preparation of this text. Zbyněk Matějů is a Czech composer who works in a variety of styles and is the ideal bridge between the musical and technical world because of his meaningful approach to composition and his hesitance to rely on tech to achieve the end results.

It is important that you get to know Zbyněk and hear his story because he has lived a colorful life in the musical world with a career slowly growing in Europe during communism after WWII and has continued for many decades across the globe. In his own words he describes his career and compositional process, with poignant insight into the relationship between sound, technology, and musical art. He originally wrote the following section in Czech which was translated into English which presented a unique challenge in communicating some fairly significant artist ideas but received his stamp of approval. More importantly, his influence is throughout the entire book and has affected the approach to explaining the tools to keep the focus on fulfilling the vision of the music. In many ways his contribution is akin to that of the frame around a beautiful painting in a museum. It shouldn't be overlooked and instead helps the viewer focus their attention on the art, while also making the art possible through structure and protection. The effect of the frame cannot be overstated.

*
**

My career began somewhat unconventionally – I was expelled from the Elementary Art School (state institution) for lack of talent. I was repulsed by the mindless piano exercises at the age of eight and preferred to improvise and play back my own ideas, sometimes variations on the compulsory piano etudes. I began to study piano privately with a professor who understood my desire and knew that if I'd continue studying music that I wouldn't be a concert pianist.

Sometime around the age of 15 I won a Czech Radio broadcast competition for young novice composers with a song for soprano and piano. While studying at the grammar school (natural science) in Rychnov n.Kn. I commuted to Prague for private composition lessons with Dr. J. Feld. It was still far from clear whether I would pursue music professionally or whether I would follow my father's footsteps at the Medical Faculty of Charles University. He was a pediatrician and my family anticipated I would seek a medical education. I also considered studying aircraft design at the Czech Technical University.

After graduating from high school, I finally decided to "take the risk" to study composition, which lasted three years at the conservatory level and five years at the Academy of Performing Arts in Prague (HAMU.) After finishing at HAMU, I had problems with the Communist regime and could not find a solid job. Thanks to my filmmaker friends I was able to compose film music and the very first film was awarded the Intervision Award which led to more opportunities. I returned to HAMU when the political shift in 1989 allowed it, and completed a three-year internship at the Department of Theory under the leadership of Dr. Karl Risinger (then equivalent to the current PhD). After completing my studies, I decided not to stay in academia where there were many former leaders of the communist regime and decided to risk things as a freelance composer.

From the beginning I have been very close to music theater and film, but not concert music. While studying I wrote a lot of concert music, mostly for my fellow instrumentalists. It was harder to get into the theatrical forms, but at school I was aware that I would like to pursue musical theater. I was approached by a student of choreography, Marcela Benoniová, for whom I composed a short ballet called Phobia for three trumpets, piano and drums. The premiere took place in Nyon, Switzerland and in Prague. At that time (1982) it was a very successful performance and I started to work with more choreographers: Jan Hartman – Vila Dei Misteri (fl. Vl.vla, piano) – prem. Brooklyn Dance Th., and M. Benoni – Pierot (solo harp and two dancers.)

In 1973 I saw Jaromil Jireš's film "Valerie and the Week of Wonders" with fantastic music by Luboš Fišer and I decided to write a ballet using the same model. Upon completion I boldly called Pavel Šmok, who at that time was probably the best choreographer who hadn't emigrated. To my surprise Pavel Šmok did not throw me out and during the very same evening when I showed him the libretto and score, he began to think about how to bring it to life. Due to the fact that it was too big of a cast for his "Prague Chamber Ballet," he devised a strategy to get the work into the most prominent theater in Prague. It never happened because neither Pavel nor I had a chance at that time against the communist authors but I was eventually able to collaborate with him after the Velvet Revolution with the Golem Ballet staged by the Prague State Opera.

During the totalitarian regime I was approached by director and screenwriter Jiří Středa and I wrote the full-length chamber opera "Anička Skřítek a slaměný Hubert," which is based on a work by Vítězslav Nezval. We applied for a grant from the Music Fund, but we were told that the work was bad, too abstract, and unsuitable for children. We decided to fight for it and finance everything ourselves. The work was presented by the Theater of Diversity in the city of Most, and the opera became very popular being performed about two hundred times. It was later screened in Poland at the Groteska Theater in Krakow. There are ten instruments used in the opera and the singers have speaking parts in addition to singing. We used a backing track for the Diversity Theater performances, with instrumental parts we recorded in a rented radio studio. After this experience I never applied for additional grants because the distribution of money was skewed towards various interest groups of the Communist authors, of which group I definitely did not belong.

During my studies at HAMU, I caused some "academic" problems by not communicating enough with the department about performing my compositions outside the school. At that time, we had a school in the Rudolfinum, which is the seat of the Czech Philharmonic. The head of the department noticed a poster on the way to school, where I had a cello sonata as part of the chamber concerts of the Czech Philharmonic. I was supposed to inform the school about my activities, but when I submitted the song and won second prize, the school learned of this from public statements. I didn't understand why I had to report everything, and so I didn't do it. The Music Fund scholarship I got in the 1980s was only symbolic, but it allowed me to choose a mentor. To the dissatisfaction of the fund management, I arranged for myself to study with Svatopluk Havelka. He was not a favorite of the regime and several of his works were banned by the Communist party, but I highly respected his compositions and especially his film music. We understood each other very well and remained close friends until his death.

I had a similar relationship with the legendary artist Adolf Born, whom I met working on the animated opera "About the Tap Who Sang at the Opera." For me it was a dream collaboration with a trio of legendary filmmakers: director Doubrava, screenwriter

Macourek and artist Born. I was given the script and total freedom regarding the casting of the orchestra and the deadline for submitting the work. As soon as I finished everything, I called the production and we recorded the music and singers and then animation started. At the time, I decided to cast singers with pop music experience and force them into something far outside of their comfort zone. It was interesting to see that pop music singers approached the work much more responsibly than opera singers. A great inspiration for me was meeting with legendary director Jan Švankmajer while working on his film Faust. Sadly, my task was only to prepare musical cuts for the operatic films. Since the death of his "court composer" Zdeněk Liška, Švankmajer has mostly used archival music. I have been working with his lifelong collaborator (sound master) for many years, more recently on the animated claymation bedtime story "Earthworms."

In my work I try to be open to all styles, genres and technologies although, for some reason, I find it natural to work with film. I think the restrictions from the previous regime blocking me from having a different career allowed me to do more work for film, where I was able to try various things within different genres and styles. I used different technologies, different instrumentation, and different sonic experimentations with music; and all of this was useful in the composition of concert music. Only once did I cause friction when I brought my synthesizer to the recording studio, with the intention that after recording my film score with the orchestra I would add more sounds to the recording. An American co-producer representative strictly refused the synthesizer, saying that I have a large symphony orchestra and they did not want to "destroy it with chemical sounds." For the film I was forced to write modern jazz interpreted by professional jazz players. It was a series about Dracula's brother-in-law, where I had the symphony orchestra with a pop singer and a howling dog in 3/4 time. I even added the sound of a storm to the mix. The picture was edited kaleidosopically with the finished music and the result was very impressive.

Working with directors can also be very frustrating for composers. On one film, where a father executes his own son in the end, I wrote a very complicated timbre composition composed of sharp clashes within instrument groups at the request of the screenwriter and producer. I had enough time for composing since I received the picture material in July and the music was to be recorded in September. I was proud of the song but unfortunately the director didn't like it and I had to write new music overnight. At the request of the director I gave it a distinctive melody and pleasant harmonies with a spectacular symphonic sound. The director was thrilled but I was ashamed and I refused further work.

A slightly different and somewhat similar situation which I experienced with a director who did not have time to record music and at the end of the recording peered over and said he had a different idea. Since everything was recorded by the orchestra, there was nothing I could do and I finished the film with the sound engineer. After some time, the director called me and told me that the music was absolutely great, he apologized and we won the award for the best TV movie of the year! The entire thing happened again with the same director on another project.

Concert music has the advantage that the composer is the only creator, apart from the role of an external text and interpreters giving the work to the audience. In contemporary "artificial" music, I feel a certain degree of being closed-off, sometimes bordering on snobbery. One example is the Paris IRCAM. Composers working there are very distinctive in style, mastering their craft perfectly, writing excellent music, perfectly built and instrumented … but I have interviewed several times with my colleague Sylvia Nicephor, a pupil of P. Boulez, saying that contemporary music is sometimes militant in nature and contrary to the freedom of art. When I listen to concerts from contemporary French music, I often find them all similar, although created by different (undoubtedly excellent) composers. I see similarities in other European music centers and I miss openness to other opinions.

To create, in my opinion, means to create new to the best of my conscience, but "new" does not mean original at all costs – to be original at any cost is self-serving, less sincere and untrue. I perceive many such works as self-purposeful, snobby looking into yourself. Unfortunately, on the other hand, there are many composers parasitizing on older music, without bring their own perspective. Such works usually fall through the sieve of time, as we can see in history, and are forgotten. I created for myself the term "kissed" music – music that "breathes" that appeals to me and I feel its depth. Unlike "unpopular" music– "music for music," although perfectly technically mastered, often impressive sounding, but without the deeper internal tension for which I listen to music.

I ponder on the term "postmodernism" – as a platform of a certain openness. It began to be used thanks to Jean-Francois Lyotard's book *La condition postmoderne* (1979). When I use the term "postmodernism," I do not mean the name of a style within a certain period of time. For me, postmodernism means openness to composi-tion, openness without prejudice. Of course, we can specify openness in the various components of the work (aleatoric procedures), but here I mean really general openness not closing in front of anything. On the other hand, I do not consider extreme (the most extreme) forms of open work – such as songs as online software – to be completely happy. Philosophically speaking, I feel that this is a total closeness to the possibility of being open as the author – the author of a work.

There are some compositions I consider to be turning points in my work in terms of orchestration. Here is an example of a title page and composition notes:

VEILING
(OBESTÍRÁNÍ)
flauto, arpa, synthesizer
ZBYNĚK MATĚJŮ

Figure 1.1

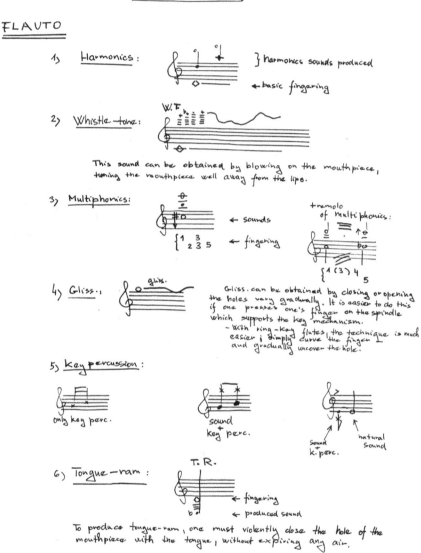

Figure 1.2

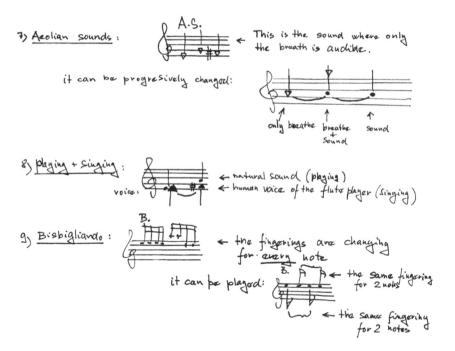

Figure 1.3

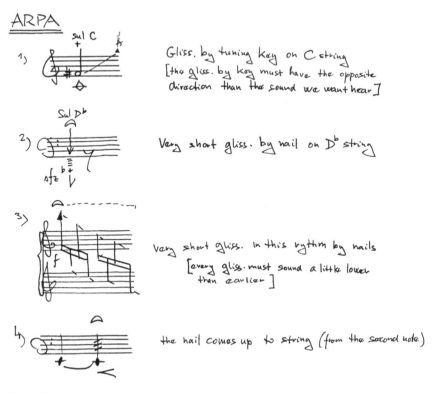

Figure 1.4

SYNTHESIZER

— it can be played by any keyboard.

The tones have to evoke impression of de-tuned little balls (or celesta etc.)
The last accord has to be played by string orchestra (very gentle) or human voices ... (ad lib.)

Figure 1.5

*) open G# key

Figure 1.6

Figure 1.7

Figure 1.8

Figure 1.9

I originally conceived of *Veiling* as a compositional study, an etude, but did not intend to share it publicly. It was created in 1993, relatively shortly after my graduation from the Music Faculty of AMU. For some time, I had been getting deeper into the sound possibilities of the flute and harp, using the latest techniques of playing both instruments. I decided to create a composition showcasing their full sound and technical ability, but also used the acoustic instruments in conjunction with an electronic synthesizer, which gave the song a somewhat isolated timbre. The piece was finally fully realized with my wife, harpist Barbara Váchalová, flutist Veronika Pudláková-Kopelentová, and me playing the synthesizer at the premiere. I used the Yamaha SY series since I did not intentionally specify the type of electronic instrument, only describing the type of sounds I had in mind. I did not want other artists to be limited by the exact type of instrument; on the contrary, I was convinced that every other performance would be slightly different in sound. Of course, I specifically chose the two above-mentioned artists for the premiere because I knew they were intensively engaged in contemporary music. (Barbara performed the Czech premiere of the song Sequenza II – for solo harp – by Luciano Beria, Veronika studied flute with, among others, Pierre-Yves Artaud, co-author of the great publication *Present Day Flutes* – the indispensable handbook for composers.)

The song was a breakthrough for me in some ways, not only using a combination of acoustic instruments in conjunction with electronics in concert music, but the entire process of fully utilizing the sound capabilities of the instruments involved.

As for my approach to composition, I usually start my initial idea in a sketch on paper. I gradually complete the sketch with details and only in the final stage transition it into its final form. I notate using Finale, which is a necessity for me because my handwritten manuscript is very difficult to read. Of course, a computer program has its limitations, and sometimes it takes me a lot of time to overcome them in terms of reaching all of my ideas without compromise.

I relate to composers who prefer to handwrite scores.

For most new compositions I initially have a somewhat visual idea of the whole shape, which is comparable to a fantastic vision of a sculpture that I am trying to materialize into some form of sound. Of course, I fully realize how binding our environment is in this process due to the places we come from, the geographical locations that define us, and the bindings of education. It is very difficult to rise above this and to overcome influential stereotypes that return one to the ruts which limit their imagination and slowly try to transform the original vision. Every new song is a battle for the final shape.

I almost never use Finale to play audio. If I want to make a sound recording and can't afford live musicians, I work with professional sound masters. Thanks to my short experience as a music director of Czech Radio and quite frequent work for film, I have many friends among the sound engineers who can help me with the score realization.

Let's return to discussing my work for Czech Radio. Since I was one of the only ones who was interested in contemporary music among music directors, I was able to be involved with fascinating sound recording projects. During the recording and live broadcasting of concerts, I worked as a music director with K. Penderecki, L. Bernstein, G. Kremer and other personalities – and, of course, with Czech composers. Set apart from traditional schooling, when you regularly hear completely new music with a score in your hand, it is a school in its own right.

In parallel with the production of concert music, I also deal with the composition of musical dramatic works (operas, ballets) and compose film music, as I mentioned previously. Working for film is also a challenge for me in terms of orchestration, if I have the opportunity to work with acoustic instruments. First, I have a chance to try out special instrumental combinations and ways of playing individual instruments, to verify how they sound naturally and possibly to revise everything on the spot. This experience can then be used to compose concert music or musical drama, where musicians expect the final version of the parts and do not like to accept any changes.

Another exciting step in the realization of film music is the subsequent mix and post-production. I choose sound masters who will be open to experimenting and willing to spend time trying different possibilities. I can give you one example from the last few years when I wrote very intimate music for one of the television films directed by Bohdan Sláma. During the finishing work, I copied the stereo harp tracks several times to other tracks and moved the timeline in milliseconds while simultaneously tuning them separately a few Hz. This created a very special section which significantly supported the film's storyline. The director watched the whole soundtrack creation process with me in the studio and it was clear that he shared the enthusiasm for the "birth" of this new music.

Recently, I have been struggling to create music accurately on my computer. Realizing the sound concepts on acoustic instruments seems less complicated than on the alternative digital versions, which are not as flexible. A necessary intermediate step for me is therefore a sketch, which is very complicated in my effort to capture the outline of all segments of the future score. I can't imagine typing straight into the computer. I would be very delayed by the computerized realization of graphic details that would correspond to the concept of the idea. I have a very close relationship to fine art and certain visual artifacts evoke sounds in me and I transform that into a sound image. I take those visuals and look for the best notation, as close as possible to traditional notation that can be decoded by live musicians. If some degree of freedom is possible, the listing may be more general.

In film music, I have learned that it is good to specify as many details as possible. If the musicians are unsure of the correct performance of the recorded music, then there is a lot of conjecture and it becomes necessary to explain everything orally and waste time in the recording studio. Thus, what can be written into notes should be written down. In my first work for film, I thought my musical ideas were obvious and easily transferable to musicians using a very general notation. However, it took me some time to explain things that seemed perfectly logical to me and so I began to define everything that was possible in the music notation from then on. As for personal reflection, this teaches us and always reveals our own deficiencies. It's a never-ending process.

In scores I try to use notation that is as close as possible to what the artist should perform in order to make the recording process require the fewest explanatory notes as possible. I used to deal with these problems using photography and tried to find alternative ways of writing various parts of the songs called "new music." I was particularly interested in the compositions of the Polish school, such as Penderecki and Serocki. Consultations with performers play an irreplaceable role when creating such scores. It is interesting to see how many parties "adapt" to their image and to better readability while respecting the original sound intent of the author. From Czech music, there is an example in the harmonic mix of the harp part of Janáček's *Taras Bulba*, and the final harp part of Smetana's *Vyšehrad* which are played a little differently so that the repeated chord tones are not muted. Speaking of the harp, there are many examples where performers add "pedalization" to the final parts of the composer. Practice has taught me that it is better not to label harps with pedal marks, but to leave this work to the players themselves. On acoustic instruments I am always attracted by the "rawness" of the sound and its unpredictability. Every instrument sounds different and each interpreter is different.

In the 1990s I received a song commission from the French Ensemble Fractal. During my stay in France I wrote the piece *Saxomania* for an outstanding French artist. It is a very difficult composition for solo saxophone, which calls for "circular" breathing. The artist does not have a chance to breathe regularly within the longer part of the score. The song was later included in a concert in Prague, but the

Figure 1.10

Czech premiere faced a problem in getting an artist to play such a difficult piece. In the end, we luckily found a jazz musician who played the saxophone and bass clarinet and he played the song fantastically. It is true that instrument technology has advanced considerably since the 1990s, and artists are facing increasingly complex challenges.

Figure 1.11

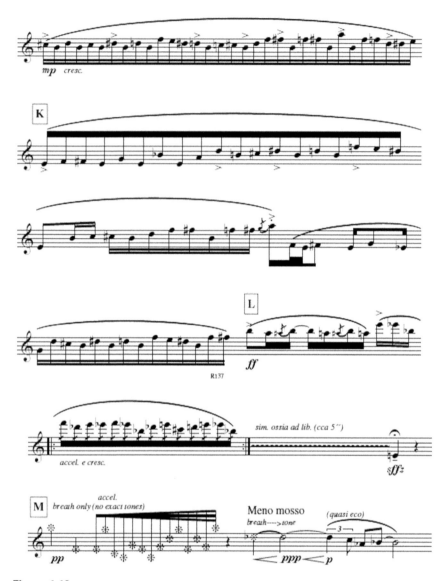

Figure 1.12

The possibilities of acoustic instruments are a big draw for all composers.
The unique sound of the acoustic bass clarinet has attracted me for a long time.
The following sample is from a solo composition titled *Tabullarium*:

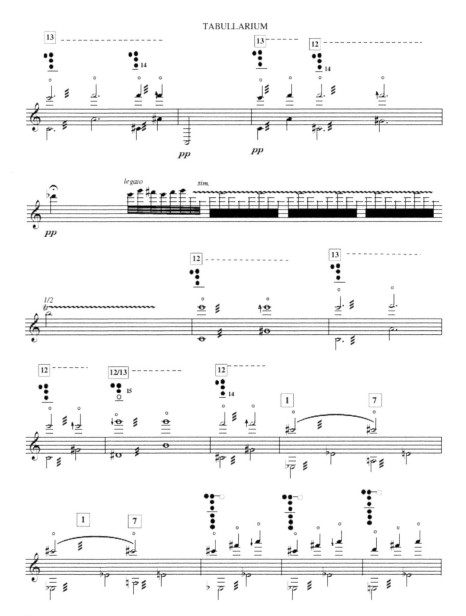

Figure 1.13

Multiphonics played by special fingerings, their trills, and slap tones form a large portion of the arsenal of this instrument. In my opinion, all this can hardly be replaced by electronic realization without losing that characteristic rawness with some imperfections, which have beautiful results.

I love working with micro-intervals, whether they arise from the use of natural harmonic tones due to the effects of tempered tuning, or if they are written intentionally in individual parts of the composition.

Here I can again present a sample of my composition *Still Life* for cello and string orchestra with vibraphone and crotali (antique saucers.) The song received a

STILL LIFE

Figure 1.14

fantastic premiere by a Chinese soloist (from memory) with outstanding orchestral players (2015, Los Angeles.)

The following showcases non-articulated whisper and timbre noise on string instruments at the beginning of the second part of *Still Life*.

STILL LIFE

Figure 1.15

In my experience I have seen how difficult it is to replace a live harp sound with a digital version. If the harp is part of an orchestra and it is about replacing short stretches, or creating a harp timbre, such as a kind of "coloring" of the overall

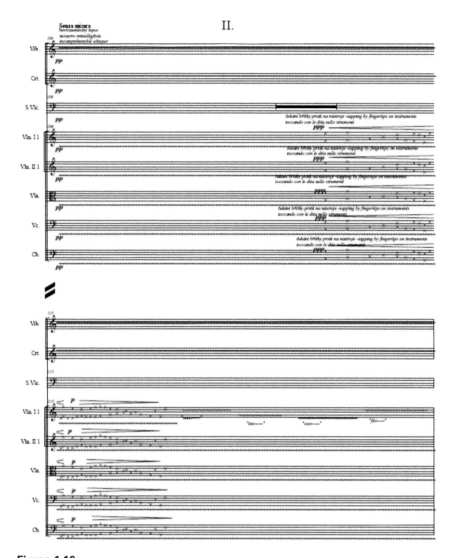

Figure 1.16

orchestral sound with a harp sound, it is not that difficult. If it is a longer part played by a harp or harps, then any digital replacement is really difficult. Unlike real sound, digitally generated sound becomes boring and drab after some time.

I enclose samples of scores from which I tried to create "demo" recordings using sample sounds, but I always humbly returned to live sound, with great respect for the instrument, capturing it with a microphone.

A similar situation occurs when combining a harp with other instruments, such as piano or celeste. Interestingly, the sound of the piano and celeste can usually be convincingly replaced digitally, but in conjunction with the harp it is very

TO-Y

pro jednu nebo dvě pedálové nebo keltské harfy

per uno o due arpe pedale o celtiche
for one or two pedal or celtic harps

Zbyněk Matějů
*1958

Figure 1.17

Figure 1.18

difficult. Random murmurs, harp pedal sounds, imperfections in attenuating adjacent strings, etc. disappear from the final sound. The listener usually does not notice this because they are part of a complex harp sound, but, in their absence, the listener may feel the overall sound to be too sterile.

Here is a sample from my score – it is the opening subtitle music for the film *Operation Silver A.*

★
★★

Figure 1.19

Key concepts

Before moving onto the rest of this text which focuses on tools and workflows, there are a few concepts which add some context to modern digital orchestration.

The blending of old and new

It is possible that previous explorations of this topic would have shown the dichotomy between analog and digital technologies, or in the context of composition it would be paper/piano vs computer/MIDI. The brightest version of the future no longer requires these different options to be separate and musicians are able to create their music using any number of tools that fit their own preferences.

Examples of what this means include tablet notation apps which let one write music as if on a piece of paper and it is nearly instantaneously converted to formal notation. Another example is of a MIDI sequencer which is attached to a hybrid piano and a musician can perform without engaging a click track and the software interprets the performance creating a tempo map of the free form recording which is used in the orchestral project. One could even sing a part into a microphone where it is easily converted into any instrument part that is desired. These are things which would've been considered witchcraft by Mozart, and yet would likely have been welcomed with open arms by composers through the ages because the distance between thought and music sound is getting shorter and more efficient than ever before.

In an age where orchestras are expensive and sometime impractical, more people have access to the same timbres and musical sounds than ever before. Yet this isn't a call to end the formation of orchestras or the communal performance of large groups of musicians – just another way to further the marriage between musical art and technology in situations where one isn't possible without the other.

The art of compromise

A violin can make sounds which are vibrant and unexpected, ever-changing and yet still familiar. An infinitely powerful computer system could potentially recreate every single possibility of what it takes a skilled musician seconds to perform. Digital orchestration is therefore a series of compromises since the tools, as advanced as they are becoming, are still limited in their ability to fully recreate what is possible in a human orchestra. For each topic that is explored and process which is explained, keep in mind that it simply isn't going to be the same, but that doesn't mean the end result won't sound realistic or be meaningful.

A compromise in this situation means that it's important to prioritize which parts of the performance are most important and to focus on making them great. In a series of audio tests shared with audio engineers, with excellent discernment, it was nearly impossible to fool them when comparing a real instrument playing and the same part performed using a sampled instrument. There are just too many small clues – when prompted to see if they could tell the difference between a real and a "fake" they turned on their hyper analysis and it was relatively easy to answer the question. The catch is that when the samples were played without a reference, or without the expectation of one being a sampled instrument, then the audio sample often passed through unnoticed, without triggering the thought of it not being a real performance. The key is to use some smoke and mirrors to make the majority of tracks sound good, but the stars of the performance sound amazing.

Mixing in the special sauce

In the days of Beethoven and long ago, students learned how to compose by copying parts and mimicking the writing of their teachers. When studying digital orchestration, there are a number of exercises which can help students master the process. One such is to mimic the sounds of actual performances. Use a combination of solo and ensemble recordings for maximum coverage, but nothing beats this as an education in orchestration.

On the other side of the same coin, in the process of learning about the instruments it becomes clear that sometimes the real thing is the best thing. In such cases 90 percent of the parts could be sampled instruments with a few key parts being recorded in the studio with actual performers. This type of combination builds realism and fools listeners into believing the entire production is real. It also has the potential to create a successful end result without breaking the bank.

The arrangement matters

As all of the tools are covered throughout this book, keep one overriding principle in the forefront of your mind: the digital tools can never surpass the quality of the music itself. A poorly written score doesn't magically transform into a masterpiece when it is realized with the most expensive orchestral library. Similarly, even a great melody can be hindered by a poorly designed arrangement. During the mix process described in Chapter 6, a lot of techniques are described but these would be nullified by an arrangement which is muddy or cluttered or even too simple. While this book doesn't look at the orchestral arrangement itself, this still has the ability to affect every part of what digital tools are capable of in the orchestration.

The scope of digital orchestration

The goal of this text is to bridge the gap between a real orchestra playing a piece of music and the same music being played by virtual instruments in a computer. Perhaps the project is a film score, a mock-up of an orchestral composition intended for sale to publishers, or it is the backing track for a pop song. BYOM (bring your own music) and this book will help you understand the tools, techniques, and finishing touches to the project in order to really sell a digital orchestra.

Chapter 2
The instruments

The traditional orchestra is full of instruments with history, colorful sounds, and a wide variety of performance requirements. It takes a lifetime to fully master any single one of them, let alone all of them. Certainly, it is different to know how the instruments work and to write out their parts as opposed to knowing how to proficiently play each one, but with digital orchestration there is an entirely different aspect to the process involving a deeper knowledge of how each instrument functions.

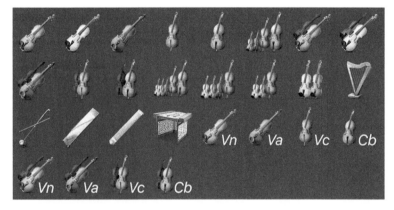

Figure 2.1 Instruments Icons in Logic Pro

There is difference between popular film scores and contemporary composition for string instruments. There is a lot of variety and there are gray areas in what can be accomplished in the orchestration process, and while it is technically possible for an instrument library to recreate any sound possible, there are no such libraries in existence and it isn't a practical proposition. The orchestral instruments are explored here to set the bar, but there are many very creative composers who think of new ways of writing for them and combining them in unexpected ways. This book is not aimed at composers who are pushing the envelope with sonic innovation and non-traditional performance techniques, but many of the topics explored still apply and will help when using virtual instruments.

In the following pages are descriptions of the instruments of the orchestra, each accompanied by a frequency analysis which shows what the instrument looks like in a way most have never seen. The results show some interesting instrument characteristics, reinforcing what someone might expect with more content in higher frequencies for the violin and lower frequencies for the tuba. The lesson here is less about expectations and more about the individualized nature of each instrument. Think of the graphics as fingerprints, showing that each instrument is something different from the next, each with their own unique timbre.

The traditional orchestra

Composer and professor Dr. Joshua Harris agreed to write about each of the instruments in the orchestra to help showcase how they work and what makes them unique. This section doesn't include fingering charts, breathing techniques, or such things as instrument maintenance information because those things aren't necessary for digital orchestration. It is possible that a fuller understanding of each instrument would help with the overall process, but most of the technical parts of the process are handled by the software instruments. The following descriptions include a lot of basic information but pay close attention to the composer's perspective. There are things which give life to each instrument and roles that each fulfill which can't be replaced. After the description of each instrument, there is an exploration of common digital instruments, common types of instrument tools, and a big picture look at how everything ties together.

Flute

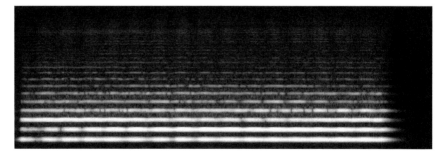

Figure 2.2 Flute

The flute is one of the earliest wind instruments, a prototype of which was played by the Greek god Pan. Blown instruments made of reeds, bamboo, and carved wooden cylinders exist in nearly every culture. The modern Western concert flute is a direct descendent of the recorder – in fact, the name of the modern instrument is technically the *transverse* (i.e., sideways) flute to distinguish it from recorders – and represents a great technological leap forward in terms of both intonation and sonic projection. Flutes are typically made of metal, with inexpensive beginner models being made with silver-plated nickel, and professional models being made with solid silver or gold.

The most common concert flute is the C flute, which plays at concert pitch and has a range of C_4 to C_6 and beyond (some models extend down to B_3). The smaller piccolo is also quite common. The piccolo sounds one octave up from its written pitch and may be made of wood or metal. Less common is the alto flute, which sounds in G, down a perfect fourth, and the bass flute, which sounds down an octave. All of these members of the flute family have more or less the same performance techniques and fingerings. They also generally play the same range of *written* pitches but transpose as described.

The flute is an edge-blown aerophone, which produces a vibrating column of air when the performer blows across an aperture on the head joint. The flute is also considered a member of the woodwind family, owing to its family resemblance with other woodwinds. The length of the tube determines the pitch of the note produced. As with other woodwinds, by covering tone holes, either directly with the keys (or fingers on a flute with open tone holes), or indirectly by pressing keys that close out-of-reach tone holes, the player can effectively change the length of the tube, changing the pitch. The angle of the air stream across the aperture and position of the embouchure will also affect the intonation of the pitch. Flutes use a very similar key system as clarinets, saxophones, and oboes, which makes it fairly easy for players to move fluently among all these instruments like speaking dialects of the same language.

The flute is notoriously inefficient in its use of the performer's oxygen. With nothing to resist the airflow, the performer must carefully control airspeed with the diaphragm and embouchure to prevent expelling all of their air too quickly, thus rendering them unable to play more than a few notes at a time. (The flute requires as much air as the tuba!) Players can fatigue from many long notes which can tax the muscles in the embouchure or long legato passages without chances to breathe. They can also fatigue from holding their instrument, especially the larger alto and bass flutes since they are longer and the player must hold the instrument with the arms a more awkward distance away from the body (this problem can be mitigated by using a u-shaped headjoint that shortens the distance from the aperture to the keys).

Wind instrument players articulate notes by *tonguing* the back of the top teeth or the gum just above the teeth, producing an unvoiced "tuh" sound. Faster articulated passages can be double-tongued, by alternately touching the back of the tongue against the soft palate, producing a "kuh" – thus, double-tonguing is effected by producing a rapid "tuh-kuh." If a more lyrical or legato tone is needed, the performer simply changes pitches with the fingers and does not articulate with the tongue. This is called *slurring*. Wind instrument players cannot slur repeated pitches, they must be tongued.

Flutists can produce vibrato by changing the airspeed with the diaphragm and throat muscles, creating a pulsing airstream that rapidly alters the intonation by a fraction of a semitone. Vibrato can give prominence to and help project the flute's tone, and is often used in solo passages. Other techniques that can change the flute's character are the *flutter-tongue* (akin to the Spanish rolled R) and the trill or tremolo (rapidly alternating between two pitches). While the flute can change registers quickly and fluidly, not all tremolos are possible or advisable, especially in the low register or when the player must depress several keys at once.

Of all the orchestral instruments, the flute has perhaps the clearest tone, closest to a pure sine wave with few overtones. It is described as crystalline or glassy, especially in the middle register. The tone is breathier in the low register, where it is also difficult to play at loud volumes. In the highest register it can be shrill or metallic because of the narrow, fast airstream required to produce those pitches. Players can get velvety, clear flute tones in the lower and higher registers by using the larger flutes (bass and alto) and the piccolo, respectively.

Flutes are well suited for very fast passages like scales, arpeggios, and other flourishes. In an ensemble, the flute, and especially the piccolo, can shine above the entire orchestra, glistening and reinforcing the highest frequencies. In solo pieces and chamber ensembles, the flute can move between a delicate lyricism and an aggressive, percussive angularity.

While the traditional repertoire for the flute has been evocative of the rustic and lyrical (remember Pan's flute), there are many effective *extended techniques* that can transform the instrument. For example, key clicks, tongue rams (forcefully stopping the air midstream by sticking the tongue into the embouchure), and *pizzicato* (a tongue click so named because it sounds a bit like a plucked string) all make the flute into a small percussion ensemble. Whistle tones, jet tones, and harmonics vary the ordinary blown flute timbre. Harmonics are produced by overblowing pitches in the lowest octave to produce higher pitches from the fingered pitch's harmonic series. Similar to the production of harmonics, a host of *multiphonics* are possible on the flute. By altering fingerings and finessing the airstream's angle across the aperture, the flutist can perform multiple pitches at once. The resulting chords are not triads or other common or classifiable harmonies, but rather somewhat random but fixed collections of pitches ranging from fairly consonant to quite dissonant, and each with a fixed dynamic range that is required to produce the multiphonic. Flutes are also quite adept at performing microtonal pitches by adopting altered fingerings, covering half of the tone hole (on instruments with open tone holes), or adjusting the embouchure. Some rare flute models have additional keys for playing microtonal pitches. Finally, *portamenti* are produced with the embouchure and are difficult for intervals greater than a minor third, except with the rare sliding head joint that allows for more controlled sliding throughout the instrument's range.

Oboe/English horn

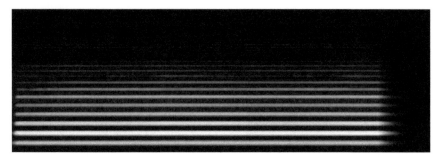

Figure 2.3 Oboe

Everyone has an opinion about the oboe – too nasal, too duck-like, too reedy. It does seem like the oboe presents the widest variations in sound quality depending both on a player's performance experience and geographic school of oboe technique. Beginning oboists' tones can be quite harsh and grating, but an advanced player can produce a lush, haunting, even angelic sound. The modern oboe, and its larger cousin the English horn, derive from the medieval shawm, a family of recorder-like instruments with double reeds, and they continue to sound somewhat at odds with the modern homogenized orchestra, like something from another time. Composers of the past have often used the oboe's unique sound to evoke the cultural other by imitating Turkish or Indian music with it. The oboe and English horn are two of the most distinctive and interesting characters in the orchestra.

Oboes and English horns employ double reeds, two reeds cut and shaved to match, then tied together to create a small tube. Double reed instruments produce sound when the player takes the reed into the mouth like a straw and blows, causing the two reeds to vibrate and set in motion a column of air vibrating in the length of the conical instrument. The length of the instrument determines the pitch of the note produced. An oboe sounds its pitches as written; it does not transpose. An English horn sounds in F, down a perfect fifth. By covering tone holes, either directly or indirectly, by pressing keys that close out-of-reach tone holes, the player can lengthen or shorten the tube, changing the pitch. The oboe's range extends from B-flat$_3$ to A$_6$. The English horn lacks the low B-flat and part of the top range, extending only from written B$_3$ to E$_6$ (but sounding down a perfect fifth). Oboes and English horns use a very similar key system as clarinets, saxophones, and flutes, which makes it fairly easy for players to move fluently among all these instruments. Most oboists double on English horn, and many orchestra pieces require this.

Because of the small size of the oboe reed aperture, it can take much longer for oboists to expend all of their air (the opposite problem of the flute!). They can typically play much longer passages without needing to take a breath. In fact, they often need to breathe before they run out of air, so they actually have to quickly finish exhaling during the break before taking in more air. The English horn reed had a slightly larger aperture but can still play rather long passages. As with other wind instruments, notes are articulated by tonguing, in this case by the tongue lightly touching the tip of the reed.

Double-tonguing is more challenging than, say, the flute, given the intrusion of the double reed into the mouth. Flutter-tonguing is also possible.

The oboe tone is often unique to the individual player because advanced players typically make their own reeds to suit themselves. Oboists in different countries also tend to make reeds a little differently, which produces regional differences in the instrument's tone. The French oboe is quite nasal and piercing and can be easily heard against a large orchestra. French composers like Debussy use the oboe to great effect as a contrasting color. The German style is smoother and warmer, matching more easily with the string instruments of the orchestra. Despite its unique, reedy timbre, the oboe is the instrument to which the entire orchestra tunes. That is because it cannot be tuned very quickly, unlike the other instruments, so it makes sense to have them tune to the oboe. In the middle and upper registers, the reediness fades and the tone becomes much clearer and lyrical. The English horn is less reedy throughout its range. Though a member of the oboe family, its physical dimensions allow for greater resonance, producing a more mellow, darker tone.

The oboe enjoys a position as a unique color, but it is quite agile, easily navigating quick passagework, scales, arpeggios, etc., including large leaps, with ease. It also can play rather loudly throughout its range with proper breath support. Playing quietly in the low register can be challenging because of the amount of air needed to activate the reed. With few exceptions, the oboe and English horn can perform trills, tremolos, and a variety of articulations. They also are capable of performing a variety of multiphonics ranging from quiet and mellow to loud and harsh to metallic and shimmering.

Clarinet

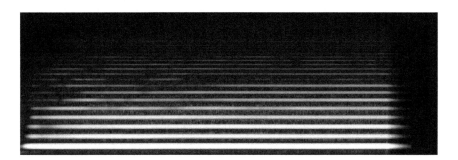

Figure 2.4 Clarinet

The clarinet is one of the most versatile instruments in the orchestra. It is found in orchestras, wind ensembles, jazz bands, klezmer groups, and other styles. In the classical and romantic periods, the clarinet was an infrequently used color instrument, and only a few concertos or chamber pieces that feature the clarinet were composed. Examples include Mozart's clarinet concerto in A major and a few of Brahms's last opuses. Clarinets are usually made of wood, although less expensive models are made of plastic or hard rubber.

The clarinet is a single-reed aerophone, meaning it produces a sound when the player blows across a single thin reed that vibrates against a mouthpiece. The resulting vibration sounds in the clarinet's cylindrical tube, producing a pitch that corresponds to the length of the instrument. Unlike other instruments in the orchestra, the clarinet produces odd partials (or overtones) more loudly than the even partials, producing a sound similar to a square wave oscillator. The resulting sound is sometimes described as hollow or woody. The clarinet has four distinct registers: the chalumeau, the throat, the clarion, and the altissimo. The chalumeau is deep, rich, and dark. The throat register includes only four pitches. It is not as full (a bit suppressed), and can challenging for the player to control the intonation. The clarion register is clear, bright, and consistently in tune. The altissimo is flute-y and delicate. The clarinet can play any dynamics throughout the entire range of the instrument.

The soprano clarinet in B-flat is the most common clarinet and has a range stretching from written E_3 to E_6 and beyond. It sounds down a major 2nd from the written pitch. The A clarinet has the same written range but sounds down a minor 3rd allowing it to play slightly lower pitches and have a warmer tone throughout its range. The bass clarinet (also in B-flat) is quite common and sounds an octave lower than the soprano clarinet – down a major 2nd plus an octave. The contrabass (in B-flat) and contralto (in E-flat) clarinets sound down a major 2nd plus two octaves and a major 6th plus two octaves, respectively. The alto clarinet is less common and less standardized, as there are alto clarinets in various keys, but usually in E-flat. The clarinet called an E-flat clarinet is not an alto but a soprano clarinet that sounds *up* a minor third. The E-flat clarinet has a distinctive, strident tone that can be easily heard in a full wind band texture or playing outdoors but can sometimes be too harsh.

The clarinet uses a key system very similar to that of the flute, oboe, and saxophone, meaning that players can easily switch between these. And all the members of the clarinet family share the same fingerings. One important feature of the clarinet is that because it is a cylinder and overblows at the perfect 12th (perfect 5th plus one octave), not the octave as most other woodwinds. Woodwind instruments are able to produce a higher register of pitches by overblowing, and this is often aided by a small hole, which is opened by a register key. On other woodwinds – the saxophone, for example – the register key causes the pitch one octave higher to sound without changing any other keys. That's why saxophonists call it the "octave key." But the clarinet's register key produces not the second partial (the octave), but the third partial (the octave plus a perfect 5th). (Remember the clarinet's overtones include only the odd partials.) This registral break on the clarinet is a little more difficult to traverse than that of most woodwind instruments, and composers must be aware of certain passages and leaps on the clarinet that cross the break (which lies between written B-flat$_4$ and B$_4$ on all clarinets) might be problematic.

The clarinet player performs articulations, like other wind instruments, by tonguing. Because the mouthpiece is inside the player's mouth, double-tonguing is generally not performed by the clarinet, although some performers are quite proficient at it. Other standard articulations including the slur are easily performed on the clarinet. The instrument can trill or perform tremolos with ease, but the composer must be aware that the clarinet player cannot perform tremolos effectively when too many keys need to be depressed at the same time, or when the tremolo crosses the registral break. The tone holes on a clarinet are open, so it is a little more difficult for the players' fingers to seal completely during a tremolo.

Unlike other woodwind instruments, the clarinet can play *portamenti* over a wide range of up to an octave or more, especially in the upper clarion and altissimo ranges. A famous example of this is the opening of *Rhapsody in Blue* by George Gershwin. In the lower registers the clarinet can only perform *glissandi* like other woodwind instruments. Vibrato can be produced either with the jaw or with the diaphragm. Other extended techniques that are possible on the clarinet include multiphonics and tongue slaps, where the player creates suction between the tongue and the reed, then pulls the tongue off, creating a pop. Tongue slaps are especially effective in the chalumeau register and on the bass clarinet.

Saxophone

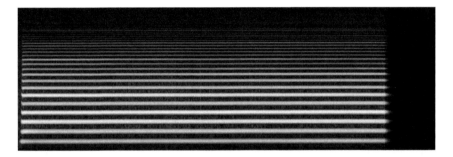

Figure 2.5 Saxophone

The saxophone family is one of the more cohesive instrument families, having a single inventor and being invented relatively recently. Adolphe Sax patented the saxophone in the 1840s and built many varieties in different keys. The most common modern saxophones are the soprano (in B-flat), the alto (in E-flat), the tenor (in B-flat), and the baritone (or "bari," in E-flat), though there are both smaller and larger instruments used less commonly. The saxophone is frequently seen in contemporary music, especially jazz, popular, and art music as a soloist and member of chamber groups. It is also a regular member of the wind ensemble and many marching bands, and though not a regular member of the orchestra, it is called for from time to time in that ensemble.

Though the saxophone is made of brass, it is classified as a woodwind since it is a single-reed aerophone, producing sound when air is blown across the reed, like the clarinet. The resulting sound is edgier than the clarinet, though, sounding odd and even partials with more balance. Depending on the player's embouchure and airspeed, the saxophone can sound very noisy, brassy, or honky, or have a focused, smooth, warm tone similar to a bowed string instrument. The saxophones have a fairly consistent timbre throughout the family – also like the strings – with overlapping pitches between the alto and tenor, for example, sounding very similar. The alto and soprano saxophones can play smoother more consistently, while the tenor and baritone saxophones, with their larger reeds, can be a bit noisier if needed.

All of the saxophones play the same written range, from B-flat$_3$ to F$_6$ and higher, though they all transpose differently. The soprano saxophone in B-flat sounds down a major 2nd from the written pitch (like the B-flat clarinet). The alto saxophone in E-flat sounds down a major 6th. The tenor and baritone saxes are in the same transposing keys, but one octave lower, with the tenor saxophone in B-flat sounding down one octave plus a major 2nd (like the bass clarinet), and the baritone saxophone sounding down one octave plus a major 6th.

The instrument uses a key system very similar to that of the flute, oboe, and the clarinet's overblown register, meaning that players can easily switch between these. And all the members of the saxophone family share the same fingerings. The saxophone's register key is known as the octave key because it overblows at the octave by design. The saxophone's registral break, between written C-sharp$_5$ and D$_5$, is much easier to traverse than the clarinet's. The saxophone's altissimo register is flute-y and can be fragile or piercing, depending on the volume, the higher it goes.

The saxophonist performs articulations like other members of the woodwind family, by tonguing. Because the mouthpiece is inside the player's mouth, double tonguing is not always performed by the saxophone, although some performers are quite proficient at it. Other standard articulations including the slur are easily performed on the saxophone. The instrument can trill or perform most tremolos with ease, but the composer must be aware of a few exceptions. Vibrato is common on the saxophone and can be produced either with the jaw or with the diaphragm.

The saxophone has very large tone holes that are opened and closed with key-controlled pads. These are well suited for audible key clicks, an effective percussive extended technique. Key clicks even have pitch content, so they sound a little like temple blocks. Other extended techniques that are possible include multiphonics and tongue slaps, where the player creates suction between the tongue and the reed, then pulls the tongue off, creating a pop. Tongue slaps are especially effective on the larger reeds of the tenor and baritone saxophones.

Bassoon

Figure 2.6 Bassoon

The bassoon, the largest double reed instrument, descended from the Renaissance instrument the dulcian. Bassoons have a distinctive bocal, a curved metal tube extending from the top of the folded conical instrument onto which the double reed is affixed. It is an immensely flexible instrument featured in a variety of musical styles and ensembles, including the orchestra, the wind ensemble, chamber ensembles, and as a solo instrument. In addition to these classical ensembles, the bassoon may be found in a bassoon quartet or choir, or in jazz and popular music, especially indie styles. Bassoonists are somewhat rare compared with other instrumentalists, in part because of its inordinate cost. A bassoon can cost upwards of $50,000 or more.

The bassoon is a double reed aerophone with a folded conical bore. The sound is produced when the player takes the reed into the mouth like a straw and blows, thereby vibrating the reed and activating a vibrating column of air inside the instrument. The bassoon has a system of tone holes and keys like the other woodwind instruments, however it has many more keys than any of the other instruments. There are multiple registers on the bassoon, each of which has a different fingering system, making the bassoon one of the more complicated woodwind instruments to play. Unlike other woodwind instruments with a single register key, the bassoon has a system of various keys that provide access to the overtone registers of the instrument. Although many woodwind players do learn to double on the bassoon, it is not as easy a transition as with most of the other woodwind instruments. The bassoon does not transpose, and the even bigger contrabassoon sounds down one octave from written pitch.

Because of the large size and heavy weight of the bassoon, and especially the contrabassoon, performers do not hold the instrument and play as they would the flute, oboe, clarinet, etc. Bassoonists may use neck straps like saxophonists or a strap that lies across their chair and upon which they sit that attaches to the bottom of the instruments and supports its weight. The bassoon can be difficult for players to control sonically, which produces a wobbly effect earning the bassoon its nickname as the "clown" of the orchestra. However, advanced players are able to produce a beautiful, velvety tone.

The bassoon has a nasal, reedy sound in its lowest register. It can also play quite loudly in that register. Its middle register contains the smoothest, darkest, most lyrical timbre. This large middle register matches well with the English horn, and complements the cello, French horn, and trombone. The normal range of the bassoon is generally considered to be from B-flat$_1$ to C$_5$, although the use of the highest notes of that range is relatively new in the history of the instrument. Until the premiere of Stravinsky's *The Rite of Spring* in 1913, whose opening bassoon solo ascended to D$_5$, those pitches were considered impractical. Now they are quite common for advanced high school and college players. Some players can go as high as G$_5$. In the highest register, the bassoon tone becomes fragile and very difficult to control.

The bassoon can perform trills and tremolos, though because of the complex fingering system, not just any tremolo is possible, especially in the lowest part of its range where fingerings are more difficult. Like other woodwinds, it can perform a variety of articulations, including flutter-tonguing and double-tonguing, with relative ease by an experienced player. The bassoon can also perform many multiphonics and microtones. The latter are possible both by changing the embouchure and by using alternate fingerings.

Some unique extended techniques are possible on the bassoon because of its many keys and larger reed. Key clicks are very effective, as is the key rattle. The reed may be removed to play special effects by itself, sounding a little like a duck call. The bocal can be removed from the instrument and attached to the reed to add some different effects. Tongue slaps, like those performed on the clarinet and saxophone, are possible on the large bassoon reed.

Horn

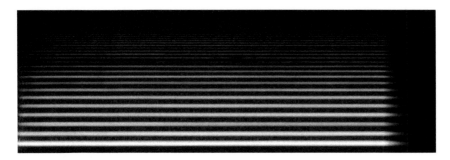

Figure 2.7 Horn

The French horn, or simply horn, is an engineering marvel. Its sound recalls its ancestors, the wooden, straight alphorn or early brass, coiled natural horn. The alphorn is around 8 feet long or so, and it was developed to communicate across Alpine valleys. The natural horn, as well as a high percentage of the classical horn repertoire, is associated with shepherds and hunters calling across open spaces or signaling to their hounds.

The horn is a brass aerophone, which means the sound is produced by vibrating the lips (i.e., buzzing) into a mouthpiece with a small conical resonator attached to a pipe. The length of the pipe determines the pitch of the horn, but that pitch, called the fundamental, is not the only pitch available. A single length of pipe, either straight or coiled, can sound the pitches of the harmonic series above the fundamental pitch. To access these harmonics, the performer must speed up the air stream and/or increase resistance of the embouchure. The modern horn is 12 to 13 feet long. The long pipe length allows for many high overtones, and in fact, in order to play scales and stepwise melodies, performers must play the above the eighth partial. The intervals between the first seven partials are wider and don't allow for stepwise playing.

Historically, natural horn players could change the fundamental pitch of their instrument by replacing part of the pipe, called a crook, with another crook of a different length. However, no matter which key crook they used, they were always limited to the pitches in the harmonic series of that fundamental pitch. The modern horn incorporates valves that route air into pipes of various lengths, effectively allowing the performer to change the "crook," and fundamental pitch, simply by depressing a combination of valves. The hornist can access stepwise pitches in the lower register by changing valves and can play higher pitches more in-tune. The modern single horn is in F, sounding a perfect 5th below the written pitch. Most modern hornists play on the double horn in F and B-flat, though for ease of transposing, the entire double horn is said to sound down a perfect 5th. Notes on the B-flat side are accessed with a thumb trigger and improve the overall intonation of the instrument. The horn's written range is from C_3 to C_6, with lower and higher pitches possible.

Because the hornist usually plays in the instrument's upper partials where pitches are very close to each other, accuracy can be quite difficult. It is not unheard of to hear even professional hornists miss their target or falter when holding the note, especially higher notes. Also, because those higher partials of the harmonic series are always a few cents sharp or flat from equal temperament, hornists are trained to compensate either by lipping up or down, or by partially stopping the bell with their hand. This is quite easy since the hand is already in the bell when performing on the horn. Stopping the bell with the hand, however, doesn't merely alter the pitch, it can change the timbre of the horn dramatically. The double horn gives the performer twice the number of harmonic series from which to choose, so players using that instrument are able to play more consistently in tune without affecting the timbre.

The physical design of the horn, with its small, conical mouthpiece resonator, long pipe, and narrow, conical bore produces a dark, almost muted timbre compared with other brass instruments. It is a powerful and focused brass sound, like a laser cutting through the orchestra. The majority of the horn's range can be played at any volume, communicating quiet confidence or aggressive power. It can sound very brassy when called for, though it isn't the usual performance practice. In the lowest register, especially, the horn can sizzle a little bit. The fundamental pitch of the horn is a special effect called a pedal tone after the pedal registal or the organ, and like organ pedals, pedal tones on the horn can be perceived as low rumbles by listeners; they can be felt as much as heard. The highest register pitches are quite loud and piercing.

The horn is found in orchestras, wind bands, chamber ensembles (including as a standard member of the woodwind quintet, despite the obvious discrepancy of category there), and even in jazz ensembles. There is a strong tradition of horn concertos and prominent horn solos in the orchestral repertoire (especially in the works of Richard Strauss, which are much-loved by hornists). Horns often find themselves used to great effect in film scores like *Star Wars* and *Inception*.

Despite its rather unwieldy mechanism and difficulty in controlling, the horn is capable of surprising dexterity, performing fast passages and leaps with ease. (This is, in part, due to the fact that leaps on the written part are never quite as far apart through *harmonic-series space* when played on the horn.) Like woodwind instruments, brass instruments articulate notes with the tongue, and since there is never a reed or mouthpiece in a bass player's mouth, flutter-tonguing, double-tonguing, and even triple-tonguing ("tuh-kuh-duh") are possible. Brass instruments can also slur, though sometimes a player must use a soft tongue ("duh" or "thuh") to assist in certain leaps, especially descending leaps.

Like all brass instruments, hornists can use various mutes to color the tone, but they can also use their hands to stop the bell to produce another muted effect. A partially stopped horn will sound flat, but a fully stopped horn will actually sound sharp. Brass instruments can also produce multiphonics, though in a very different way than woodwinds do. Brass players play one pitch while singing another, and if the two pitches are perfectly consonant, a third pitch can sometimes be head. Special effect *portamenti* – called rips – are idiomatic to the horn. They differ from *glissandi*, which are produced by fingering a chromatic scale rapidly. Trills and tremolos can be performed with the keys and can also be performed with the lips.

Finally, because everything old is new again, several contemporary works for horn call for natural horn technique – that is, making use of a single harmonic series and hand position in the bell to perform. Composers may also be seeking "out-of-tune" microtones either for special effect or for just intonation. This technique, which used to be the only technique available, is somewhat specialized now – like the novelty of driving a manual transmission car with a clutch in an age where most cars have automatic transmissions.

Trumpet

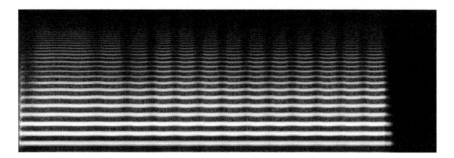

Figure 2.8 Trumpet

The trumpet shares a great deal in common with the horn and other brass instruments in terms of technology, mechanical construction, and sound production. However, its origins as a military instrument give it a different kind of auditory significance. Historically the trumpet's immediate ancestor the bugle hailed the king, called the troops to advance or retreat, and signaled the times of day (as it still does on military bases and scout camps). The bugle, like the horn, was used to communicate across wide spaces, but unlike the horn, its cylindrical bore, relatively short pipe, and larger, cupped mouthpiece give it a brighter, edgier tone.

The modern trumpet, like the bugle (which is also called a natural trumpet), is a brass aerophone with a cylindrical bore (the cornet is very similar to the trumpet except that it has a conical bore and warmer tone). Its pipe length is just under five feet, and it has three piston valves that, in various combinations, change the pipe length and fundamental pitch, giving the player the pitches from seven harmonic series. The trumpet can play all the chromatic pitches from G-flat$_3$ to C$_6$ and beyond. The pedal tones are not usually played as the tone is quite fuzzy or airy, but they can be used as a special effect. Some professional players can play an entire octave higher, to C$_7$, often with the aid of special mouthpieces. The trumpet and cornet in B-flat are common in wind ensembles, jazz bands, and marching bands; they sound down a major 2nd from their written pitches. The trumpet in C, more common in orchestras and has a slightly brighter tone, plays the same written pitches but does not transpose.

Like other brass instruments, the trumpet can play multiple partials from the same harmonic series by only changing airspeed and embouchure. Thus, to play higher pitches, the player uses faster air and a more resistant embouchure (players and conductors try to avoid using the phrase "tighter lips" in order to prevent a variety of performance problems resulting from inadvertently tightening the rest of the body, but they're tightening their lips). Since they usually avoid the first partial, trumpeters typically play the second through the eighth partials, except for the out-of-tune seventh partial. Some pitches' intonation, especially when using the third valve, must be compensated for by adjusting tuning slides. Experienced players do this automatically.

Like any wind instrument, the trumpeter can only play for as long as they have breath to play with. The trumpet provides moderate resistance to airflow, allowing the player to play most phrases without needing to breathe. Long notes or a long time spent in the upper register can cause embouchure fatigue. This lip fatigue can be quite severe at times, causing the player to need a long time to recover, even overnight.

There is a general distinction between the timbres of conical brass instruments, which tend to be darker, and cylindrical brass instruments, which tend to be brighter. The horn and the cornet are conical, so dark and mellow. The trumpet is cylindrical, so a bit harsher, or *brassier*, by comparison. A brighter sound indicates the presence of additional higher frequencies in the spectrum, which is caused by material, mouthpiece resonator, pipe length, and bore shape and width. The trumpet can play very softly or very loudly throughout the lower two thirds of its range. As it goes higher, it becomes more difficult to play softly. The lowest register can sound a little buzzy as vibration of the lips is slower, allowing us to hear the discrete iterations of the buzz. As the trumpeter ascends into the middle and upper registers, however, that sense of discrete vibrations is lost, and the tones are much smoother.

Trumpets can perform a variety of tongued articulations, including various accents, slurs, double-tonguing, triple-tonguing, and flutter-tonguing. Trills and tremolos are possible, with combinations of fingers and lips depending on register, especially if only one valve needs to be changed. Some trills can be awkward. Trumpets can perform with vibrato, as well.

Trombone

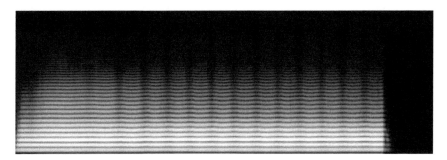

Figure 2.9 Trombone

The trombone has the simplest mechanism of all the brass instruments and has been around the longest in a similar form, since at least the 16th century. In English, the early trombone was called the *sackbut*, though when the modern instrument gained popularity as a member of marching bands in the 19th century, the Italian *trombone* was preferred. Trombones are standard members of the orchestra, wind ensemble, marching band, jazz band, and brass chamber ensembles. The trombone also figures prominently in some popular styles like funk and ska.

The trombone is a brass aerophone, played by vibrating the lips into a resonating, cup-shaped mouthpiece, and with pitch being determined by the pipe length. It is transparently a single length of pipe, though its length can be quickly changed by a free-moving slide. The standard performance practice divides the slide's length into seven approximately equidistant positions that correspond to seven semitones. Each position gives the performer access to the harmonic series of that position's fundamental pitch. Higher partials tend to be more out of tune, but that's easily compensated for by adjusting the slide position. When higher partials are available in more than one position, the player tends to prefer the closer positions in order to simplify performance. Some trombones have an F attachment that, when activated via a thumb trigger, allows them to access a second pipe, coiled ahead of the slide, that essentially gives the player access to all the pitches in fewer positions.

The tenor trombone is by far the most common, though the bass, alto, and valve trombones are used regularly. Trombones do not transpose, but they are sometimes named for the fundamental pitch in first position. The tenor and bass trombones are said to be in B-flat, not because they transpose, but because the first position sounds a B-flat. The alto trombone is in E-flat (but again, it doesn't transpose from written pitch). The tenor trombone's common playable range is from E_2 to F_5 and beyond. The bass trombone is also quite common, and its range extends down to B-flat$_0$.

Like other wind instruments, the trombone can articulate notes with the tongue, including double-, triple-, and flutter-tonguing. Slurring presents more of a challenge, since the pipe length cannot be immediately altered, but must always slide. Trombonists will usually perform a softer, so-called legato, tongue ("duh" or "thuh") to hide the slide and approximate a slurred articulation. The trombone, like other brass instruments, can play very loud throughout its range. It can also play quite soft except for the highest part of its range.

One look at the trombone and you know what makes it special. The slide that dominates the instrument's design makes it the only wind instrument that can play a consistently controlled *portamento* anywhere over its range. Though it can play a continuous *portamento* anywhere, the pitches available to the slide are not indefinite. The trombone can slide between any pitch in the B-flat harmonic series (i.e., 1st position) and a tritone away at most (i.e., to 7th position). Pitches from other harmonic series are limited as to how far down they can slide since they cannot make use of the complete length of the slide, which is 7 positions or six half steps. Triggers on the tenor and bass trombone allow for tritone-long slides beginning on more pitches, and longer, special-effect slides might allow the player to reset the slide quickly in the middle of a long slide to extend the effect. The slide also allows for any microtone be performed or for any intonation system to be utilized. Like other brass, mutes can color and dampen the timbre.

Tuba

Figure 2.10 Tuba

The tuba is the lowest brass instrument, and it comes in a variety of sizes, shapes, and pitch ranges. The baritone horn, also called a euphonium, is a smaller kind of tuba. Some tubas have the bells pointing upward, some have the bells facing forwards. There are marching tubas that sit on the player's shoulder, and the sousaphone has the pipe wrap around the player's torso and shoulder. Tubas generally have a conical bore, though there are variations. There are also different kinds of valves including piston, rotary, and compensating valves which will be explained below. There's usually one tuba in an orchestra, though sometimes there are more. Tubas, as the bass instrument of the brass family, are found in brass quintets and choirs as well. They are common in marching bands and in jazz ensembles, especially Dixieland combos.

Like the trombone, tubas are named by the open fundamental pitch, though they do not transpose. In other words, though various instruments may have different open pipe fundamental pitches (e.g., B-flat, F, E-flat, C), they play the pitches as written. Contrabass tubas sound even an octave lower (BB-flat, CC, etc., where the double letters refer to an old-fashioned low octave designation). Surprisingly, differently pitched tubas have different fingerings, apparently to preserve a standard pattern of valve combinations across all valved brass instruments, including the horn and trumpet. This pattern of seven valve combinations lowers the fundamental pitch by one semitone for each combination, and it corresponds to the seven slide positions on the trombone. Tubists can generally navigate pretty fluently between tubas. The standard tuba range is D_1 to F_4 and higher. Some tubas can descend down to B-flat$_0$.

The tuba is a brass aerophone, played by vibrating the lips into a resonating, cup-shaped mouthpiece, and with pitch being determined by the pipe length. Like other valved brass instruments, the tubist can access different pipe lengths by playing different valve combinations. As with the trumpet, when multiple valves are depressed, especially when one is the third valve, the pitch tends to be sharp. Trumpeters can extend their tuning slide to slightly lower the pitch. Because the tuba is so much lower and larger, it is not as simple as adjusting a tuning slide. Compensating tubas have a fourth (and sometimes a fifth and sixth) valve to allow the instrument to be more in tune over its entire range. These extra valves may be in parallel with the other valves, or lower on the instrument and played with the left hand (instead of the right, which plays the main valves).

Like other wind instruments, the tuba articulates notes with the tongue. The tuba can play very loud throughout its range. It can also play quite soft except for the highest part of its range, but even there it can play surprisingly soft. Though it is a bass instrument, the tube is surprisingly lithe and flexible. It can play quick rhythms and melodies in the tenor range easily. There is no reason to stick to slow, plodding bass lines with the tuba. Despite the large size of the bell, like other brass instruments, tubas can use mutes – large ones – to color and dampen the timbre.

Harp

Figure 2.11 Harp

The harp dates back over 5000 years and endures as a symbol for music. Harps, lyres, and zithers are found in many cultures throughout the world. All of them have a bank of tuned strings that are plucked with the finger (or, sometimes, played with a plectrum or hammer). Zithers (including dulcimers and autoharps) have a thin resonating soundboard parallel to the strings, while harps are open, with a resonator at one end of the strings. The modern concert harp, or pedal harp, was invented at the turn of the 18th century in southern Germany, though it has been updated a few times since then. Another common harp is the lever harp (often called the Celtic harp).

Harps are plucked chordophones. The pedal harp has a large triangular frame made of metal and wood. The player sits with one side, the sound board, leaning on the right shoulder. The strings extend from the soundboard below to the tuning pins at the top of the instrument. Though the harp is said to be fully chromatic, it actually only has seven strings per octave, one for each letter-named pitch. The harp is tuned to the C-flat major scale, but each *pitch class* can be raised one or two semitones via a set of seven pedals, each pedal corresponding to one pitch class. Changing the pedal raises the pitch of all the strings of that pitch class (i.e., the G pedal affects all the Gs on the instrument, for example). Thus, while any of the chromatic pitches is possible on the pedal harp, only one variety of each lettered pitch is possible at any time. The harp can be tuned to any major or minor scale (including the harmonic minor, but not the melodic minor since pitches cannot be altered quickly while playing). Post-tonal music can be performed if the composer sticks to the same seven pitch classes within any passage and keeps in mind the necessary pedal changes required to incorporate other pitches. Sometimes the harpist can invent creative solutions to pedaling like respelling pitches en-harmonically. The harp's range extends from C-flat$_1$ to G-flat$_7$ when all the strings are flat. The two lowest strings, C-flat$_1$ and D-flat$_1$ cannot be changed with the pedals, though they can be tuned via the tuning pin. The full range is C-flat$_1$ to G-sharp$_7$ with the limited access to pitches as described above.

The lever, or Celtic, harp is similar but smaller than the pedal harp. The lever harp, as its name implies, alters the pitch of the string by use of a lever at the top of the instrument near the tuning pin. Each lever has only two positions. Most lever harpists tune the strings to the E-flat major scale, which gives them access to most of the common keys in the Celtic or folk music usually performed on lever harps. However, the strings of lever harp can each be tuned independently. In other words, the play could be able to play a G in one octave and a G-flat in another, which is not possible on the pedal harp.

Playing the harp on the right shoulder, the right arm can't reach the lowest strings, which are farthest away. Both hands can play most of the range of the instrument, but only the left hand can reach the lower third. The diatonic glissando is perhaps the most idiomatic gesture on the harp, though the harp can play a variety of textures including melodies, melodies plus accompaniment, block chords, and arpeggios. Like the piano, the harpist will be most successful with no more than three notes per hand when playing block chords. Scales and arpeggios with more notes will be broken into smaller groups to be played by each hand.

The harp's nylon strings (some wound in metal) give it a bright, resonant sound. The low strings – steel wrapped nylon – can be quite full-spectrumed and must be used with care in ensembles to avoid muddying the middle and lower registers. The highest strings carry very well and, in ensembles, give a reinforced attack to the flute or violin.

A common effect to reduce the resonance of the strings – akin to a muted violin – is to play near the soundboard. This technique is called *près de la table* and produces a sound like a classical guitar. The harpist may also approximate a xylophone (*sons xylophoniques*) by muting the string with one hand and plucking with the other. There are many other extended techniques that alter the timbre of the harp: plucking the string with the nails, the thunder effect (*glissando*-ing very loudly on the low, metal strings), hitting the bass strings for a percussive "whomp," knocking on the soundboard, sliding a hand up and down the bass strings producing a whistling sound, bending the pitch with a tuning key, and *preparing* the strings by weaving paper in them to create a rattle when plucked.

Piano

Figure 2.12 Piano

primary visual metaphor for most of Western music theory. But it's a relatively new instrument – its modern form was only finalized in the late 19th century – and it's still evolving. Of course, keyboard instruments have a much longer history, with instruments like the harpsichord, clavichord, virginal, and organ reaching back to the medieval era and earlier. Keyboard instruments are also found in musical cultures all over the world, as well, including the African marimba and the Indonesian gamelan.

The mechanism unique to the piano is a felt-covered hammer that strikes a string (or group of two or three strings tuned to the same pitch) when the key is struck. The earliest of these hammered instruments was the *fortepiano* (also sometimes called the *pianoforte*, for maximum confusion), which was invented at the beginning of the 18th century. It was called the *fortepiano* (Italian for *loud-soft*) because the performer could control the volume level by playing the keys harder or softer. This contrasted with keyboard instruments of the time like the harpsichord that could not change dynamics. The piano steadily improved with technological and engineering advances over the decades. The repertoire from Mozart to Beethoven to Chopin to Liszt to Rachmaninoff reflects an evolving instrument. The keys got deeper and heavier, the hammer action became more consistent and responsive, the range expanded, the strings got thicker, the sound board and resonating wooden case improved, and the instrument became louder and more brilliant. Recent improvements include an extended lower range on some instruments and the incorporation of MIDI on some for communicating with digital instruments and computers.

The piano is generally considered a percussion instrument since the primary performance practice is striking the keys with the fingers. However, it is also a struck chordophone in the zither family since the vibrating strings are the sounding bodies and the sounding board is parallel to the strings. The standard range of the piano is greater than any orchestral instrument, with 88 keys extending from A_0 to C_8. Some pianos have as many as 108 keys, with extensions at the top and bottom of the range, but this is not very common. The timbre is fairly homogeneous throughout the range of the piano. The highest pitches are thin and plinky, and the lowest pitches are resonant and boomy. This results from the string sizes required for the highest and the lowest pitches.

The main parts of the instruments that contribute to the sound are the keys, the hammers, and the dampers. Each key – one for each chromatic pitch – has one hammer and one damper. In its resting state, the damper rests against the string so that it cannot vibrate. When a key is pressed, the damper comes off the string, and the hammer strikes. The string vibrates freely until the key is released, and the damper returns to silence the sting again.

The two common form factors of pianos today are upright and grand. The upright piano, suitable for small spaces and rehearsal, has a vertical soundboard and strings, giving it a fairly small footprint. The grand piano – which has a larger footprint and is suitable for larger spaces, performances, and recordings – has a horizontal soundboard and strings. Both the upright and the grand piano have three pedals. The *damper* (or *sustain*) pedal on the right is the most commonly used pedal. It holds all the dampers off the strings so that all the strings may vibrate freely. This gives a lush, sustained sound during performance. Pianists will usually change the damper pedals (that is, release and re-depress the pedal) to clear the ringing sounds whenever the harmony changes. Otherwise, the new harmony might be clouded by the previous harmony causing an unwanted dissonance. With all the dampers removed, all the strings in the harmonic series of the played keys will vibrate in sympathy. So, the damper pedal can increase the volume of the instrument. The *una corda* (one string) pedal on the left is also called the *soft* pedal. It shifts the hammers to strike only one or two strings. The piano can still play quite loudly with this pedal depressed, but the tone is more mellow. The middle pedal is not as standardized as the other two on upright pianos, but on grand pianos it is always a *sostenuto* pedal. The *sostenuto* pedal holds up the dampers of any keys that are being held when the pedal is depressed. The performer may continue to play with a dry or *staccato* touch on all the other keys, but when they strike a key that was being held when the *sostenuto* pedal was depressed, that pitch will ring.

Pianists can play multiple notes and voices at a time, but they are generally limited to ten notes at a time (clusters are the exception; see below). Even so, it isn't possible to play ten-note chords that rapidly change; most piano music sticks to three-to-four notes per hand, at most. Another physical limitation is hand size. Even though the keyboard has a nearly 8-octave span, each hand can only reach a little more than one octave. Harmonic ninths and tenths are sometimes possible for players with large hands, but they are usually broken (arpeggiated). Large intervals usually preclude the possibility of playing notes between the outer notes with the same hand – the contortion of the hand required to reach a ninth or tenth makes it difficult to reach notes in between. Experienced pianists can play very fast passages, including wide leaps and octaves, with ease, especially the scales and arpeggios associated with tonal music. Atonal chords and melodic passages may be less idiomatic, depending on the player's particular experience.

The piano affords composers and performers many other ways to make sounds through a host of extended techniques. These are usually divided into three categories, playing on the keys, playing on the strings, and playing other parts of the piano as an idiophone. On-the-keys techniques include the *glissando* (on the white keys, the black keys, or both – though a *glissando* on both white and black keys is less effective, and nearly impossible, owing to them being on different planes), clusters (white keys, black keys, or both; the fist is approximately a fifth, the palm with fingers extended is approximately an octave, and the forearm is approximately two to two and half octaves), and various overtone and resonance effects using silently depressed keys in conjunction with ordinary playing. When playing inside the piano, on the strings, the player may perform a chromatic *glissando* on the strings (or other scales and arpeggios by holding down certain keys or using the sostenuto pedal), plucking the strings (with the finger, the nail, or a plectrum), or scraping lengthwise on the lower strings. Players can also set the strings in motion using electromagnetic devices (like the ebow), bowing with thin wire or fishing line, striking them directly with percussion mallets, playing another instrument loudly into the piano, or using other items inside. With any of these techniques, the damper pedal should be depressed. The sound envelope will change with the material used, but the timbre will sound like a piano. Finally, the timbre and overtones of the piano can be altered by inserting various materials into the strings, then playing on the keys or directly on the strings. This is called *preparing* a piano, and the resulting performance practice is called *prepared piano*.

Organ

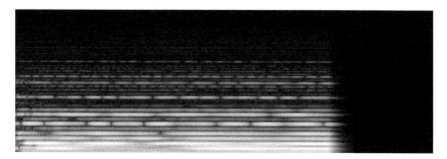

Figure 2.13 Organ

Mozart called the organ the "king" of instruments, and it is one of the oldest of the modern instruments. The earliest organs date to the 1st or 2nd century BCE in Greece, when they were powered by water. They were used secularly, not unlike the organs in baseball stadiums today. The modern pipe organ, powered by pressurized air, has not changed much in terms of structure since the 13th or 14th centuries. At that time, organs were primarily associated with the liturgical music of Catholic church. Later, in the 19th and 20th centuries, the organ became important in the classical tradition. In the second half of the 20th century, small electronic organs became a mainstay in rock and popular music.

Volumes can be written on the organ, and they have. It is very complex, and no two organs are quite the same. Hector Berlioz, in his seminal 19th century treatise on instrumentation, spent a few pages discussing specific organs in small towns across Europe. However, pipe organs all work on the same principle, namely that a set of manuals (keyboards) and a pedalboard (keyboard for the feet) control pressurized air flowing through a set of tuned pipes. Thus, an organ is an aerophone controlled by keyboards and pedalboards built into a console at which the organist sits. The physical choreography required to perform at the organ is really quite intense. The hands play multiple keyboards, pull and push stops, and activate preset *registrations*. The feet play the bass line on the pedals, open and close the swell box, and activate additional presets by the pedals. Organ performance is a full-bodied piece of movement art. It isn't uncommon to have an assistant pushing and pulling stops, and in the days before electric air compressors were available, assistants had to physically operate a bellows to keep the air pressurized.

Stops are mechanisms, usually a pull/push knob but sometimes another kind of switch on electronic organs (they're called *drawbars* on Hammond organs), that activate a collection of pipes called *ranks*. Most pipe organs have several ranks of pipes with different timbres. Most ranks have one pipe for each key on its corresponding manual or pedal. When a stop is pulled out (or otherwise switched on), that rank of pipes is activated and air will enter the pipe when a key is pressed. Each stop or rank (they're often used interchangeably) has one pipe for each key, except for mixtures. Mixtures are just what they sound like, combinations of multiple ranks. Ranks are grouped into *divisions*, and each manual or pedalboard controls one division. The organist can pull out multiple stops to activate multiple ranks (i.e., multiple timbres). This combination of tone colors, called *registration*, gives the organ incredible control over both color and volume (the more stops that are pulled, the louder it gets, until the organist *pulls out all the stops*). Some divisions of the organ are enclosed in a wooden case called a *swell box*. The swell box can be opened and closed via a pedal to control the dynamic level of that division. Not all divisions are enclosed.

It might be surprising to learn that the organ is a transposing instrument, or at many of the stops are. Stops are usually defined by a name that denotes the tone color and a number that specifies the length of the pipe in sonic feet (the measurement doesn't necessarily align with the standard unit of length of the same name). An 8' (eight-foot) stop sounds its written pitch (i.e., it doesn't transpose). Stops of 4' and 2', and, less commonly, 1' are shorter and will sound up one octave, two octaves, and three octaves, respectively. Longer pipes of 16', 32', and, less commonly, 64' sound down one octave, two octaves, and three octaves, respectively. The written range of the manuals is typically C_2 to C_7, and the pedalboard is C_2 to E_4. A written C_4, depending on the transposition of a particular stop, could sound anywhere from C_1 to C_7.

The whole point of the organ is to give one player access to the equivalent of an entire orchestra. That's why there are so many stops. Stops fall into four main categories of tone colors: principals, flutes, strings, and reeds. Principal stops are strong, middle-of-the-road organ stops. If you played the "pipe organ" patch on a cheap digital keyboard, it would sound a lot like a principal stop. The flute stops are airy and light. The string stops are warmer than the principals, but more full-bodied than the flutes. The reed stops tend to be the loudest. Though they usually sound more like a nasal oboe or bassoon, stop names like *trombone* and *trumpet* indicate that these stops are intended to approximate brass instruments. There are two other non-exclusive categories of stops: *mixtures* and *mutations*. Mixture stops combine multiple ranks, usually including higher octaves and fifths, into one stop. Mutations are stops that sound upper, non-octave partials. For example, a 2 ⅔' pipe is called a *twelfth* because it sounds a twelfth above the written pitch. Mutations are used in conjunction with other stops to color registrations.

Modern organs often incorporate electronics or digital control to some extent. In some cases, the keys control electrical switches that open and close pipes (replacing earlier systems of mechanical action that involved thousands of small wooden rods that connected keys to the pipes). In others the organ console might play electronically synthesized or digitally sampled stops. Some pipe organs have MIDI functionality. Many smaller electronic organs, like the Hammond, are common in rock, jazz, and other popular styles.

Guitar

Figure 2.14 Guitar

The guitar is a surprisingly resilient instrument, being continually reborn into new incarnations with different technologies and different abilities, and every new iteration seems to usher in a new style of music. Part of its endurability is that it's relatively small and portable. Troubadours (medieval singer-songwriters in western Europe) could carry its chordophonic ancestors as they traveled. The guitar itself originated in Spain by the 12th century and has long been associated with Spanish musical styles like *Flamenco*. Guitars have also figured somewhat prominently in the classical music tradition, with many composers writing solo and chamber works for the instrument. Though it tends to be too quiet to compete with the full orchestra, there are concertos for the guitar, usually with string orchestra. With the advent of jazz in the early 20th century, the guitar lost some ground to the banjo, which could play louder in Dixieland groups. But with the invention of the electric guitar, the banjo quickly faded, and the jazz guitar was born. The guitar was very common in the music of Latin America and spread north into American country and western styles. The guitar is associated with almost every popular style of the 20th century, including jazz, blues, country, rhythm and blues, and rock.

The guitar is a chordophone with six strings and a long-fretted fingerboard. Acoustic guitars (including the classical, Flamenco, and modern acoustic guitars) have a large resonating body made of wood with one hole directly beneath the strings (resonator guitars like the Dobro are acoustic with a metal resonator). Classical guitars use nylon strings, while modern acoustics and electrics use steel strings. Electric guitars can be hollow-bodied, solid-bodied, or a hybrid of the two. Solid-body electric guitars can be quite heavy, so they are usually cut away. This cutaway also gives the electric guitarist access to the higher frets that are more difficult on the acoustic (many modern acoustics also feature cutaway bodies). Because electric guitars lack resonators, they are very quiet when not amplified. Of course, they are intended to be amplified with built in transducers, called *pickups*, and wiring to allow the guitar to be connected to an amplifier. Some guitars have preamps, and some have digital audio interfaces built in so they can be plugged into a computer (e.g., via a USB cable), bypassing the analog amplifier.

The flat fingerboard allows guitarists to easily play chords with all six strings, and this is the most common performance practice in many styles. The guitar can also play intricate and fast melodic passages, however, as seen in the classical repertoire. In rock music both practices are used in a complementary way, with the chordal technique called rhythm guitar and the melodic technique called lead guitar. Some guitar styles use a pick (also called a *plectrum*) to strum the chords or play individual notes, while others use a style called finger picking. Classical guitarists generally have long fingernails that act as finger picks. Some acoustic guitarists use actual plastic finger picks, but most modern fingerpickers that use amplification (either an electric guitar or an amplified acoustic guitar) just use their fingertips.

Guitar strings are tuned in perfect fourths, with a major third between the fourth and fifth strings, and it transposes, sounding down one octave from the written pitch. The written open strings are, from lowest to highest, E_3, A_3, D_4, G_4, B_4, and E_5, and it can play as high as D_6. The layout of the strings allows for certain chord voicings that can be moved up and down the fretboard with the index finger laid across all six strings, acting like a moveable nut. These are called *barre chords*, and there are a few barre chords for each major and minor triad, and various seventh chords and other extended harmonies common to jazz. The voicings that fit the hand naturally on the guitar are more open-voiced – that is, with larger intervals – than the close-voiced chords that fit the hand of the pianist. Chords that use the actual nut of the guitar, without needing the index finger, are generally easier to perform, especially on the acoustic guitar, and are called *open chords*. Sometimes, in order to access open chords in a different place on the fretboard, the guitarist may use a *capo* on a fret as a moveable nut. This effectively transposes the instrument. Rock music employs a subset of common barre chords that only takes two or three strings. Called a "five chord," or a "power chord," these include only the root and fifth of the chord, leaving out the third. It's an ambiguous harmony that allows mostly modal rock music to float between major and minor.

Guitarists are famous for chasing particular sound colors, and they have many tools at their disposal in their pursuit. It is difficult to make a clean distinction between standard techniques and extended techniques, especially with the electric guitar, because players are always experimenting, creating, and standardizing new techniques. With only the acoustic guitar, a player can get timbral variety by finger-picking, strumming with a pick, or using a slide. The slide, which comes from the blues tradition, gives an expected *portamento*, but also changes the frequency spectrum with combination tones, giving a shimmering effect to the regular sound. This effect is amplified on resonator guitars, where slides are more common. On a clean electric guitar (that is, a simple amplified guitar without other effects), every nuance in performance is amplified, and yet, without the resonance of the acoustic guitar, the clean electric guitar can sound rather dull and tinny. Fortunately, there are options for changing the sound of the guitar in some very radical ways.

The first element in the electric guitar's timbre is the pickup and body style. The most common pickups are single coil (most associated with Fenders) and humbuckers (most associated with Gibsons). The single coil gives a thinner sound that dissipates quickly but is more malleable with electronic effects. The humbucker produces a warmer, more sustained tone. A hollow-body or semi-hollow-body guitar will also be a little warmer. Solid body guitars may lack some of that warmth, but they have more timbral flexibility with the use of amplifiers and pedals. Amplifiers – or amps – are frequently housed in the same cabinet as one or more loudspeakers. Technically, this is a combo amp (since it *combines* the amplifier and the speaker), as distinct from separate heads (amplifiers) and cabinets (speakers). There are many kinds of amplifiers, each with a unique sound – vacuum tube, solid state, and hybrids are common. Modeling amplifiers are becoming more common because they have the ability to reproduce a variety of other amplifier sounds digitally. Modeling amplifiers are usually played through speakers with a flat frequency response in order to preserve the digitally modeled sound.

Electric guitars and amps can produce a large variety of timbres. Many amps come with effects like overdrive distortion (literally turning up the gain on the amplifier until the sound distorts), reverb (like an echo), and tremolo (vibrato). And if that's not enough, there are thousands of pedals designed to give the guitarist that perfect sound. In addition to distortion, reverb, and tremolo pedals, some of the more common ones are phaser, flanger, chorus, delay, compression, and wah-wah pedals. Through combining different guitars, pedals, and amps, guitarists have access to nearly infinite timbres. Modeling amps and computer applications can also digitally reproduce the sound of many pedals, so nearly any of these timbres can be achieved through digital means.

In addition to the performance practice described above, there are some more experimental techniques that guitarists use. On the acoustic guitar, the player can tap the body of the instrument or prepare the strings with paper for a rattle (like the harp or piano). All guitars allow for natural and artificial harmonics, although the latter are more difficult because of the longer strings. Guitarists often use the term "chimes" for natural harmonics because of the bell-like quality of the sound. Harmonics sustain longer on electric guitars. Electric guitars can achieve noisy effects by producing feedback, which occurs when the pickups are close enough to the amp to create a feedback loop. Feedback noise is usually in the harmonic series of the stopped strings. Some electric guitarists can set their strings in motion with a bow or an *ebow*, which is an electromagnetic device that causes the string to vibrate without being touched. Finally, for the guitarist who is still looking for that perfect sound, there is no end to the possibilities provided by tinkering with electronics. Hacking can create either a cleaner signal or a noisier one, depending on the player's preference. One player's perfect sound might be another's glitch music.

Percussion

Figure 2.15 Timpani

Percussion instruments, as a somewhat standardized collection of instruments in Western musical culture, are an inherently diverse group. Percussion players find there are a few, but only a few, generalized techniques that apply to the group as a whole. Even mallet technique for instruments that are played with handheld sticks can be different from instrument to instrument. Likewise, the mechanism of sound production varies wildly among this group of instruments that is only loosely held together by the notion of striking them to make a sound. The group includes the piano and the jawbone, the windchimes and drums, the glockenspiel and the woodblock. Some cymbals, for example, are played with mallets (e.g., suspended or ride cymbals), some are banged together manually (e.g., crash cymbals), and some are banged together with a pedal (e.g., hi-hat). Some percussion instruments have definite pitches (e.g., xylophone), some are considered unpitched (e.g., snare drum), and some have approximate or relative pitches (e.g., tom-toms, woodblock, agogo bells). Additionally, there are instruments of Western origin (e.g., timpani, triangle), non-Western instruments that have been adopted into the Western mainstream (e.g., cymbals, marimba), non-Western instruments used infrequently, usually by percussionists and adventurous composers (e.g., slit drum, doumbek), and modern hybrid instruments (e.g., waterphone, steel pan). We will focus on some of the more common instruments by sound production category.

Drums is often a metonymic stand-in for percussion-as-a-whole, or just a sign that a person doesn't realize the depth and diversity of percussion instruments. Drums are membranophones, instruments that produce a sound from a vibrating of a natural or synthetic skin head pulled tightly over a circular frame. (Some non-percussion instruments have resonant membranes, like the banjo.) There are many kinds of drums, but there are a few very common ones.

The snare drum has a head on both sides of the wooden or synthetic frame. The bottom head has a series of steel wires – called snares – that touch it and vibrate when the top head is struck, producing a short buzz. The snare drum is played with drumsticks, wooden sticks with small beads carved at the head. The snare drummer can achieve a sustained white noise sound by playing very rapidly, even bouncing the sticks to play even faster. This is called a *roll*, and it can be performed to varying effects on any stick- or mallet-struck percussion instruments. The tom-tom, or simply tom, is similar to a snare drum without snares, and sometimes without a bottom head. Toms come in a variety of circumferences and depths that correspond to their relative pitch.

Most drums have an indefinite pitch, but some, like the timpani, or kettle drums, can be tuned to specific pitches. Timpani usually come in groups of two to six so the player can play multiple pitches (often the bass line) and harmonize with the ensemble. Timpani are usually played with felt-covered mallets that minimize the attack of each stroke, but they can also be played with harder, wooden mallets, for greater articulation. Bass drums are large and of a low but indefinite pitch. They can articulate slower rhythms or metric accents, or they can effect a rumble by rolling with large, soft mallets. Drum rolls on the timpani and bass drum are single-stroke rolls – that is, they do not involve bouncing the mallets on the head as in the snare and tom rolls.

Idiophones are instruments that produce sound when the main body of the instruments vibrates. Most are made of either wood or metal, and they can run the gamut from non-pitched to approximately pitched to pitched. Common wooden, non-pitched idiophones include the woodblock and vibraslap. The woodblock gives a short, dry, clicking or knocking sound depending on the size of the woodblock. The vibraslap is designed to mimic the rattling sound of the jawbone. Temple blocks are a bank of resonant woodblocks of various sizes, making hollow woody sounds of approximate pitches.

The xylophone and marimba are pitched wooden idiophones. Both have tuned wooden blocks with resonator tubes underneath, arranged as a piano keyboard. The marimba has variable-width blocks and the xylophone has fixed-width blocks. The former has more resonant, fuller timbre than the dry, crisp xylophone. Both can be played with yarn covered mallets of varying hardnesses. The xylophone can be played with hard rubber or plastic mallets for greater articulation. The xylophone sounds one octave higher than written, with a standard written range of F_3 to C_7. The marimba sounds as written, with a range anywhere from C_3–C_7 to C_2–C_7, depending on the instrument.

Common metal, non-pitched idiophones include the cymbal, tam-tam, triangle, and brake drum. The cymbal and the tam-tam both have a full-spectrum wash of metallic sound. The cymbal tends to have a sharper attack, although when a suspended cymbal is played with a soft mallet, the attack is muted. (The tam-tam is often mistakenly called a gong, but gongs are pitched, and not as common.) The triangle makes a shimmering metallic tinkle, and the brake drum, a dull thud combined with a bright clink. And anvil produces a similar sound, but is significantly heavier than a brake drum. All of the instruments above come in various sizes that change the sounds made, but they aren't normally thought of as even relative pitched instruments, though they certainly could be. Relative-pitched metallic idiophones include multiple cowbells, agogo bells, and the windchimes, although the latter is performed as a single instrument, making it more of a special effect than an instrument that approximates pitch.

Pitched metallic idiophones include the glockenspiel, vibraphone, crotales, and chimes. The glockenspiel is usually played with hard, brass mallets, and produces a piercing, loud tone. It sounds two octaves higher than written and has a written range of F_3 to C_6. The vibraphone has more mellow sound and is usually played with yarn-covered mallets. Vibraphones have resonators with electric-powered spinners inside to give the tone a variable-speed vibrato (hence its name). It also has a damper pedal like a piano that allows the fixed-width metal bar to sustain indefinitely. Vibraphones sound as written, with a standard range of C_3 to C_6, though there are variations depending on the manufacturer and model. Crotales, also called antique cymbals, look like finger cymbals, although they are heavier and usually arranged in the chromatic scale as a keyboard. They are played with hard mallets and produce a high, resonant sound that is less harsh than the glockenspiel. They can also be struck together like finger cymbals. Crotales sound two octaves higher than written, with an available written range of C_4 to C_6. Chimes are also known as tubular bells because of their long tubular shape. They hang vertically, suspended in a frame with a large damper, and are struck with a hard rubber or leather mallet. The damper is removed with a pedal to allow for the bell-like chimes to ring. They sound as written, with a range from C_4 to F_5.

There are many conventional and unconventional ways to play percussion instruments, and percussionists are among the most experimental players, especially given the modernist bent to most of their repertoire (outside the traditional orchestral repertoire). Many percussion instruments, especially metal ones like vibraphones, chimes, crotales, cymbals, and tam-tams, but even the marimba, can be bowed to produce haunting sustained tones. Others like the triangle or smaller cymbals can be lowered into water immediately after being struck to bend the pitch. Coins and other items can be taped to drum heads or sounding bodies to produce buzzes. A player can produce vibrato on the marimba by making a silent "wah-wah" with the mouth directly over the sounding bar. The list goes on and on. Anything that makes a sound is fair game on percussion instruments.

Strings

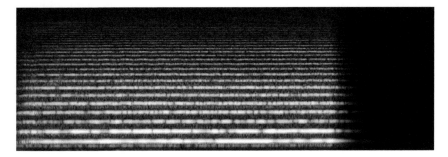

Figure 2.16 Violin Section

The violin family, including all the members of the modern string section, is the winner of natural selection. It evolved in the 16th century as a separate species called the *viola da braccio* (viola of the arm) at a time when the *viola da gamba* (viola of the leg), or *viol*, was quite popular. They existed together for some time, then the latter eventually succumbed to extinction. (Although viols are played occasionally, usually in the context of historical performance, they are rarely seen in orchestral scores since Bach's death in 1750.)

The two families of string instruments are similar in the sense that they are both played by drawing a bow of horsehair across a tuned string. Viols have flat backs, sloped shoulders, frets, gut strings, and five to seven strings tuned in fourths with a third in the middle (like a guitar). Viol players also bow with an underhanded, or German, technique. Violins, on the other hand, have curved backs, rounded shoulders, four strings tuned in fifths (except for the double bass), and no frets. The violin (from the Italian *violino*, or small viol) and the viola are also played on the shoulder, whereas all the viols were played on the lap or between the knees like a cello. Modern cellos, like the violin and viola, are played with the overhand French bow technique, while double bass (also called the contrabass, string bass, or simply bass) players are divided between the French and the German bow techniques. These differences in design and performance practice made the violin family brighter in timber, louder, more resonant, and more agile. A single violin, viola, cello, or bass can easily project its sound throughout a concert hall, competing with a full orchestra when necessary.

All the violin family instruments are chordophones and work according to the same principle – the vibrating string, or monochord. The monochord has been the exemplar of music theory and acoustics since at least the ancient Greeks. Like the vibrating column of air found in wind instruments, the vibrating string of a string instrument has a pitch that corresponds to the length, thickness, and tension of the string. When a finger stops the string against the fingerboard, it shortens the string, and a higher pitch is produced. Strings are set in vibration by two primary methods, bowing and plucking. First, the strings can be bowed. This, called *arco*, is the most basic performance practice of the string instruments. Second, players can pluck the string with their fingertips. This, called *pizzicato*, produces a short, dry, somewhat percussive sound akin to a guitar.

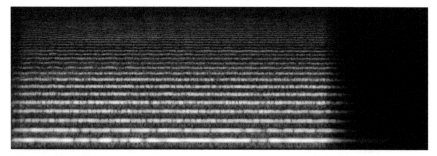

Figure 2.17 Viola Section

In addition to the stopped strings, players can access the harmonic series of each string by lightly touching fractional *nodes* along the length of the string that correspond to certain partials (but not pressing the string against the fingerboard). For example, the node at the midpoint of the string produces the second partial, the node at one-third the length of the string produces the third partial, the one-fourth node, the fourth partial and so on. The sounds produced by lightly touching the string are called *natural harmonics*.

The bodies of the modern string instruments are generally proportional with one exception. The viola has been something of an engineering challenge for centuries. Were it to have the same proportions as the violin, it would be too long to hold on the shoulder but too short to comfortably play between the legs like a viol or cello. To achieve the proper alto range, one fifth below the violin, the fingerboard is shorter, but the body is bigger. This gives the strings the right pitches, but because the strings are a bit looser, it somewhat compromises the viola's ability to match the brilliant, bright sound of the violin or cello. However, composers – especially from the 20th century on – have been charmed by the viola's uniquely smooth, muted color, turning a perceived weakness into a feature. Bartok's *Concerto for Viola*, for example, showcases the instrument's beauty and soft lyricism, a welcome contrast and counterweight to the mountains of violin and cello concertos.

Not to over-generalize, but the string section is the most homogeneous consort in the orchestra. From the lowest pitch of the bass to the highest of the violin, the section might be thought of as one composite instrument, they are so similar in timbre, articulation, volume, and so on. The biggest timbral variations of the ordinary bowed strings lie between the stopped notes and the open strings. Open strings are fuller and more resonant than stopped strings; the latter lack the resonance of open strings as they are ever-so-slightly dampened by the fleshy finger. Composers have made good use of the contrast between open and stopped strings by playing both at the same time or alternating unison pitches, open on one string and stopped on another. Notes that lie higher on the string are slightly less resonant since the string is shorter, so high notes tend to not resonate as well as low notes (although there are many factors involved, and it isn't always the case that higher notes don't resonate well). The pitch G_3, for example, would sound different when played on the open G_3 string on the violin, stopped on the C_3 string on the viola (about one-third up the string), and stopped on the C_2 string on the cello (about two-thirds up the string). The violin's open strings are G_3, D_4, A_4, and E_5. The viola shares three of those; its open strings are C_3, G_3, D_4, and A_4. The cello's open strings lie one octave below the viola's at C_2, G_2, D_3, and A_3. The double bass, which sounds one octave lower than written, has open strings tuned in fourths, with *written* pitches E_2, A_2, D_3, and G_3.

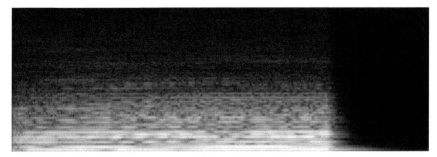

Figure 2.18 Cello Section

The layout of the strings across an arched bridge near the tailpiece makes it possible to bow two strings at once without touching the other strings. This is generally called a double stop, even though one or both of the strings may be open. Triple stops are also possible at louder dynamics by pressing the middle string down to the same plane as the two outer strings. Many triple, and all quadruple, stops are actually performed as broken chords even if they aren't notated as such. It's simply impossible to play them otherwise. Not every multi-stop is possible since the pitches must lie on adjacent strings close enough for the hand to reach both.

Pizzicato is usually performed with the right hand, which is also holding the bow, so the player uses the index finger only. The player cannot typically play pizzicato very quickly unless they put the bow down. Open strings may be plucked by the left hand. The player can bow the string in both directions, and while the so-called up-bow and down-bow have slightly different sonic qualities, experienced players work to make them sound the same. There are many *arco* techniques, usually divided between *detaché* and slurred. *Detaché* simply means that each note takes one bow stroke, so that notes in sequence are played up-down-up-down, etc. Slurred notes are all taken together with one bow stroke, either up or down (not unlike the slurring of wind instruments). Additionally, bowing techniques are informally divided into on-the-string and off-the-string techniques. On-the-string playing means the bow remains in contact with the string as it changes directions. Off-the-bow means the bow comes off the strings with a little lilt or bounce, causing both a weightier attack and a more resonant ending of each note. Some off-the-bow techniques, like the *ricochet*, play multiple notes in the same bow direction. Composers and players often combine these techniques to give passages more interest.

String players can control the frequency spectrum of their instruments through a few standard techniques. One way is through using mutes. All the string instruments can quickly place a mute on the bridge of the instrument to deaden the upper overtones of the sound. They do not reduce the volume, *per se*, but they alter the timbre, giving a slightly duller, more plain color. The natural harmonics described above effectively do the opposite by removing the lower frequencies from the spectrum, leaving only the higher frequencies behind, producing a thin, ethereal color. Only a few natural harmonics are possible on any given string – specifically, the second, third, fourth, and fifth partials (resulting in an octave, an octave plus a fifth, two octaves, and two octaves plus a major third above the open string, respectively). Higher partials are possible on the cello and bass because of their longer strings. For example, finding the seventh partial on a violin string is nearly impossible with average sized fingers because the node is so small and unforgiving. *Artificial* harmonics allow the performer to achieve the harmonic color nearly anywhere on the instrument by stopping the string, then lightly touching the string, a minor third, major third, fourth, or fifth above. This results in a harmonic pitch two octaves plus a perfect 5th, two octaves plus a major 3rd, two octaves, or one octave plus a perfect 5th, respectively, above the stopped pitch. Players can also control the frequency spectrum of the sound by bowing nearer to the fingerboard to produce a warmer tone (*sul tasto*) or nearer to the bridge to produce a brighter tone (*sul ponticello*).

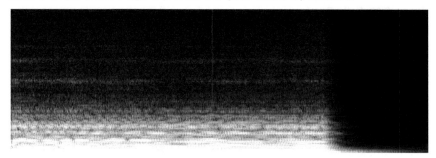

Figure 2.19 Double Bass Section

The strings are the dominant force in the orchestra and strongly signify a musical tradition spanning centuries in Western culture. The standard string quartet contains two violins, a viola, and a cello, though contemporary composers have used non-standard quartets. There are extensive repertoires of solo works for each of them, including many concertos for each, although the violin and cello have more than the viola or bass. This enduring tradition of string music in a variety of forms arises from the multipurpose facility of the instruments. They can play just about anything, melodic, harmonic, percussive, rhythmic. They can perform a variety of styles, textures, timbres, and volumes. At the same time, they seem to share a consciousness like a single organism, acting as a unit.

In addition to a historical wealth of styles and techniques collectively known as classical music, strings are put to use in nearly every contemporary popular style as well, including rock, r&b, country, hip hop, Broadway, film scores, and jazz. In popular styles dominated by rhythm instruments like guitar, drums, and keyboards, strings often serve a limited function as pads, which are long, sustained notes or chords that act like glue to connect the other rhythmic elements. Composers and producers in these styles also use strings to play melodies in unison or octaves, as well as pizzicato for textural and timbral contrast, especially in film and Broadway scores.

Over the past 200 years composers and performers have developed a hefty catalog of extended techniques for string instruments. *Col legno* bowing means to bow the string with the wooden part of the bow; *col legno battuto* means to tap the string with the wooden part of the bow. The snap *pizzicato* (sometimes called the Bartok *pizzicato*) sees the performer pluck the string away from the instrument so that it rebounds and audibly strikes the fingerboard. More recent extended techniques include overpressure (using such heavy bow pressure that the note distorts), bowing on the bridge and other parts of the instrument, tapping the instrument with the hand, and bowing on the tailpiece.

Joshua Harris is an award-winning composer and a faculty member at Sweet Briar College in central Virginia. His music, which has been performed in concert halls around the world and appeared in feature films, is grounded in a fascination with visual art, textures, sound spectra, non-linear narratives, and extreme temporal manipulations and has been heavily influenced by studio techniques of electro-acoustic composers.

Orchestral instrument libraries

It's not as simple as buying a virtual orchestra and playing all of the notes with a MIDI keyboard. There is a wide variety of instruments available and each of them has many features in common and sometimes things which are unique. It would be difficult to create a complete database of all instrument sets and their features, and it would likely be outdated by the time this is published. Instead the information is organized into categories to help digest the possibilities of what can be accomplished with orchestral instrument libraries.

Categories
The variety of instruments can be generally organized into a few categories which can help sort through a lot of options. In broad strokes there are sampled instruments and modeled instruments. Sampled instruments use audio recordings of actual instruments, which creates a very realistic version that can be triggered by MIDI and performed with very good results. This is why the sampled instrument option is the most widely used instrument type. The modeled instrument type uses a set of parameters to generate an entirely new sound which doesn't rely on previously recorded audio recordings. This

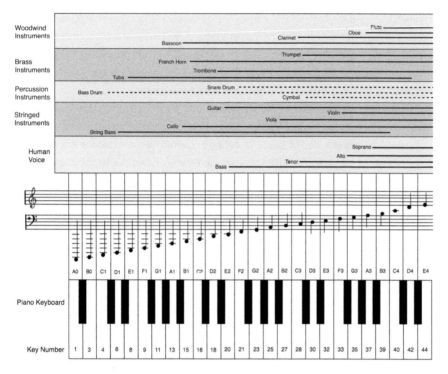

Figure 2.20

method holds a lot of promise for the future because, if designed well enough, they could potentially have more flexibility and increased power.

Sub-categories

Instrument libraries offer a plethora of different playing styles and typical timbres. The goal is to give the end user the ability to create a realistic orchestral sound by offering enough options to mimic accurate sounds. The categories include articulations, types of vibrato, changing dynamics, solo vs sectional parts, and special effects. One of the biggest issues facing orchestral libraries is the dichotomy between simplicity/usability and comprehensive instrument coverage. It isn't possible to have both a simple instrument and an instrument with maximum options.

Scenarios

In order to provide a slightly unconventional approach to describing the available instruments, the following series of scenarios are designed to highlight the features of standard orchestral libraries.

Scenario 1: quick mock-up

If the goal is to test out a composition before sending it to an orchestra for a human performance, then the requirements for an instrument library are much different than if it needs to sound like the real thing right out of the gate. If a quick listen is required then

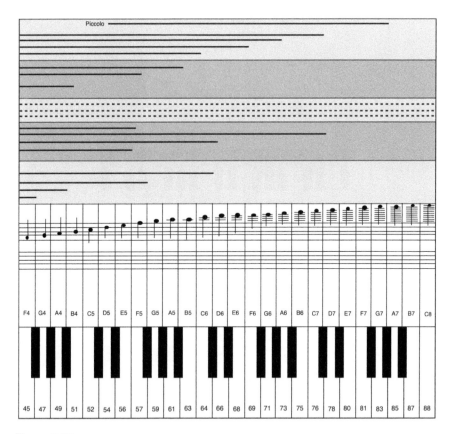

Figure 2.21

Figure 2.22 Play Instrument

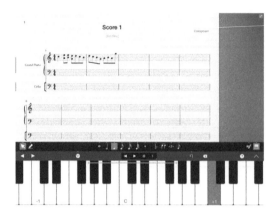

Figure 2.23 Symphony Pro App

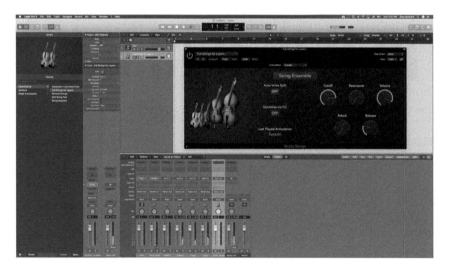

Figure 2.24 Studio Strings in Logic Pro

using the sound engine of the notation software is likely the best option, especially considering it would be able to reproduce the sound with all of the visual tempo markings, articulation markings, and things such as dynamics.

If a more accurate result is required, perhaps for a film producer who wants to ensure the score is on track or perhaps when shopping a score for potential work, then using a dedicated instrument library might be a better option in a digital audio work-station with a full featured sequencer. Such an instrument might have fewer instrument options than the biggest collections but should still be capable of high-quality sounds. The key element is to use a library which produces desired results with the least amount of effort, in terms of dynamics control, articulations, and sound selections.

PRACTICAL APPLICATION

Logic Pro X has a built-in instrument called Studio Strings which has a variety of articulations, easy to control dynamics using any MIDI controller, and the ability to tie all of these things to the notes in the score editor, including the ability to use markings in the

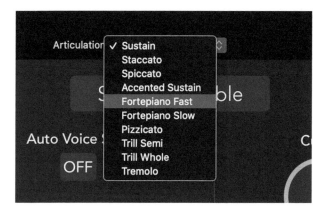

Figure 2.25 Examples of Articulations

score as a control source. Another benefit is that this particular instrument has four combo patches representing different types of orchestras which combine the individual instruments. One feature that isn't found in most other instruments is the ability to change incoming MIDI data so that a simple root position triad is re-voiced so that the chord is better in line with an orchestral voicing. This means a part can be played in a piano style with orchestra style results. This is extremely powerful because it allows for good results with minimal time investment.

Scenario 2: stylized sound

Figure 2.26 Spitfire Website

Perhaps the project is a short film without a significant budget and yet the director insists it sound like a Hans Zimmer score. Harmony and melodies aside, obtaining access to an orchestra in the studio to achieve the Zimmer sound might be out of the realm of possibility ... except that Spitfire Audio collaborated with Hans Zimmer to release an orchestral bank which imitates his "sound." Composer collaborations like this aren't unique and you can find other examples of instrument designers seeking the holy grail of sounds from well-known composers.

Whether it is the sound of a composer, the sound of a well-known orchestra, or even the sound of a well-known recording space, choosing a library based on a specific sound is a valid way to decide. Perhaps it makes more sense to add specialty orchestras to the arsenal as an expansion to the core instrument, but it really depends on the type of projects they'll be used for.

PRACTICAL APPLICATION

The Spitfire Audio Hans Zimmer instrument was recorded with a specific group of musicians, in a specific recording place, and with a specific set of performance styles. The Spitfire library is built on the Kontakt sampling engine, which offers a lot of power and flexibility. The instrument touts having a compact and streamlined user interface which means that the project can be accomplished with minimal different instruments.

Scenario 3: traditional orchestration with realism

Figure 2.27 East West Browser

In a situation where the end goal is producing the sound of a traditional orchestra with a high level or realism then there are a few instrument options which are more suited for the job. Vienna Strings is one of the most comprehensive libraries with thousands of samples, offering many articulations and more playing styles than most. It also offers multiple instrument placements, with reverb that accurately replicates the best concert halls. But there isn't a simple interface which easily handles every single aspect of the project and it takes time to program all of the parts.

Another set of instruments that has a comprehensive set of sounds are the libraries from East West. These are perhaps slightly less comprehensive than the Vienna sets, but still an excellent option for orchestrating realistic end tracks. Each instrument in the orchestra is replicated with typical variances, but also with extremely detailed options such as alternate fingerings, harmonics, and various tunings.

Programming with these instruments can be daunting because of the sheer number of options and so it takes practice, time, and serious skill.

PRACTICAL APPLICATION

Working in Logic Pro with the East West Hollywood Strings is a strong combination

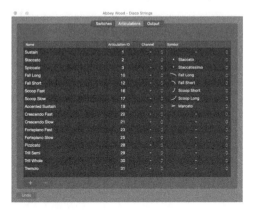

Figure 2.28 Articulation ID Dialog

because, when sequencing, both the instrument and Logic have tools which work together. The strings have many different articulations which can be played one at a time or triggered by other MIDI inputs in a process called key-switching. Logic can be used with its own process of assigning Articulation IDs to switch between the different sounds. As a result, each note can trigger the pitch, loudness, and articulation simultaneously which makes the orchestration easier to create and more efficient in the required data.

Scenario 4: Larger than life

In this situation the project could easily be a film score with bombastic drums, lush strings, intense brass, and a few exotic solo instruments. It's rare that a single library has everything from both traditional instruments and everything else, but sometimes

Figure 2.29 Logic Pro Instruments

the same company sells multiple libraries to cover everything imaginable. The East West Composer Cloud is an example of a bundle with a huge range of instruments from nearly every continent and culture. Combining styles can be very impactful but a single library isn't likely to adaptable to the needs of different genres. In some cases, the workstation also has instruments which add to the toolkit. Logic was mentioned earlier and has one of the largest set of sounds and instruments, but Cubase and Pro Tools also have instruments with a variety of additional offerings.

One of the most difficult parts of combining sources from widely different places is making sure they end up sounding like they belong together. It is also quite difficult to know how each instrument is supposed to sound and be performed, which means that the less you know about the instrument then the easier it is to use it inappropriately or outside of what it is actually capable of in the music. Never use an instrument you don't understand unless you are willing to consult with someone who does before completing the project.

PRACTICAL APPLICATION

Figure 2.30 East West Reverb

There are several specific tactics to use when combining multiple instruments to help create a cohesive end result. The most important is to know when to use the internal effects on the individual instruments as opposed to avoiding them for the effects in the workstation. Instruments are often promoted because of audio effects which are bundled within the interface and can be used solely with the instrument but not with other instruments. If the entire instrumentation could be created inside a single instrument, then this might be a really good benefit but not as much when combining.

When creating a cohesive experience such as putting an ensemble into a virtual concert space, it is unnecessarily complex to create individual spaces around each instrument separately and is much more efficient to use a single tool to create the space around all of the instruments.

Scenario 5: Synthesizers
Not every combination is between digitally created acoustic instruments, with a strong option to being layered with a synthesizer or otherwise electronically created

instrument. These instruments have endless sonic possibilities. chapter 3 explores the role of synthesizers in significantly more depth.

Figure 2.31 Hollywood Strings

East West library example

There isn't as much depth in this section covering the available libraries, which isn't by accident. Libraries come and go, with new bells and whistles announced every month. There are smart instruments which can sing words you type for them, and string sections which can do the work for you. When you invest into one of the major libraries it is a commitment of time and practice to get to know how it works and what is at its heart.

Content

The first step is to get to know what is in the library. Which instruments are available? What features is the library known for? The example in the section is the Hollywood Strings Diamond library, which is a significant instrument in the orchestral department. There are so many instrument choices that it can get really difficult to know which is the right one to pick. For instance, do you know when you should use a violin section with or without vibrato? The Hollywood Strings has violin instruments which specialize in

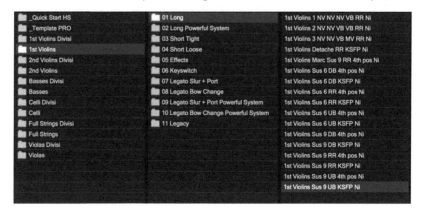

Figure 2.32 Finger Position Patch

various things. One set is programmed so that you can control the finger positions. Another set has multiple versions of each note, which is called round robin, thus creating an instrument which can play repetitive sounds without sounding repetitive. Another instrument uses the mod wheel to move through various note lengths so that as notes play, they can become longer or shorter over time, just as a real performer could easily do. The list goes on.

Lingo

The library is full of codes to help figure out which patch to load. That's the easy part and all it requires is a reading of the manual. The harder part is knowing when and how to move between the various instruments. Is it even possible that a bank with 50+ violin patches can even have sonic continuity between all of them? In other words, when the instrument switches from one to the next does it sound like the same orchestra? In theory it should have perfect continuity because it was the same orchestra recorded in the same recording studio, but, in reality, there are enough factors which prevent it from always being a smooth transition.

Trial and error

The Play instrument is capable of loading multiple patches simultaneously, which can then be triggered using key-switches. Once a collection of patches is created then the

Figure 2.33 Multiple Instruments Loaded

question becomes if they will actually work together. It's easy to switch between them, but if it doesn't sound seamless then it won't work, and alternatives should be explored. The smallest variations can throw a listener and make the transition obvious. The best libraries are carefully crafted to avoid variances, and this is certainly why inexpensive libraries are often less functional.

The ultimate compromise

The more specific the musical concept in the mind of the composer or orchestrator, the less likely a library will ever meet their expectations. Testing shows that comparing an instrument library performance with the real thing is stacked against the library. However, if the library is able to stand on its own then there is a much better chance of sounding realistic. The key is to compromise what is heard in your head as opposed to what the library is able to produce. It is better to go with what works than what sounds perfect. This is a hard compromise to make for a composer, and it is clear from years of experience that many aren't willing to even consider anything less than their personal vision. If this hurdle can be crossed, then it is possible to achieve amazing results that sound as good as the orchestra's full potential. In fact, here lies the heart of the issue ... think of working with an instrument library as it being a living, breathing orchestra – arranging for it should include taking into consideration the abilities of that orchestra. Write what it is capable of and then you'll never be disappointed in the results.

Chapter 3

Synthesis in orchestration

Since the earliest days of orchestral composition, the possibility of creating and hearing full orchestrations without needing a full orchestra of musicians has been an appealing, yet mostly unattainable, concept. As studio technology advanced and musicians began experimenting with different orchestrations on their songs, the need for instantaneous access to a multitude of different instruments became crucial.

Early instruments

In this section the foundation is explored for the technologies which set the stage for digital orchestrations to ever exist. At first the technology was rudimentary in comparison to where it has arrived, but at the time it was revolutionary.

Mellotron

One of the earliest instruments capable of recreating instrument sounds was known as the Mellotron. Developed in England in the early 1960s, the Mellotron was arguably the first sample-based instrument. It utilized a number of pre-recorded tapes that were triggered by pressing notes on a piano-type keyboard. By speeding up and slowing down the tape, the Mellotron could achieve various pitches. By incorporating individual tapes for each key, full polyphony (the ability to play multiple notes at once) was achievable. The Mellotron contained short recordings (about eight seconds) of various instruments (strings, flutes, etc.).

Figure 3.1 Mellotron

These recordings were stored on tape and could be played back and looped whenever a key was pressed. This innovative instrument allowed musicians to add instrument sounds to their songs without needing to hire session musicians to come in and play. This type of flexibility opened up a whole new world of experimentation into the recording studio.

Early adopters of the Mellotron included bands such as The Beatles, The Rolling Stones, Pink Floyd, and Led Zeppelin just to name a few. The Mellotron began to be seen as its own unique instrument with its own unique sound due to the unique mannerisms and quirks of magnetic tape playback. Due to the fact that the Mellotron's sound could not be easily manipulated and customized, it wasn't long before musicians and composers started looking elsewhere for ways to mimic orchestral sounds. The next evolution would come in the form of the electronic sound synthesizer.

Moog instruments

Recreating orchestral instruments on synthesizers is a practice almost as old as synthesizers themselves. In fact, one of the earliest commercial appeals of the Moog Modular systems was the possibility to mimic acoustic instrument sounds without having to hire numerous musicians. The promise of being able to create an entire orchestra with one

Figure 3.2 Software Mellotron

**Figure 3.3 The Original Moog Opus #1 - Housed in the Stearns Musical Instrument
Archive at the University of Michigan**

machine made these early modulars an attractive investment for many studios despite their typically prohibitive cost. Although recreating sounds that already existed went against the thoughts and dreams of early synth manufacturers like Robert Moog, Herb Deutsch, and Don Buchla, it was hard to deny the sonic potential these machines held for instrument recreation.

The synthesizer as we know it today was arguably invented in the small quiet town of Trumansburg New York in 1964. Although other people were experimenting and creating synthesizer predecessors at the time, Robert Moog is most often credited as the synthesizer's inventor. Dr. Robert Moog, an inventor and owner of the R. A. Moog company which made small amplifiers and Theremins, was working with composer Herb Deutsch on what would become known as the synthesizer. Deutsch was an experimental composer who was striving for sounds that did not exist. After meeting Moog and talking about Theremins and the possibility of a device which could satisfy Deutsch's quest for otherworldly sounds, the two began a fruitful partnership.

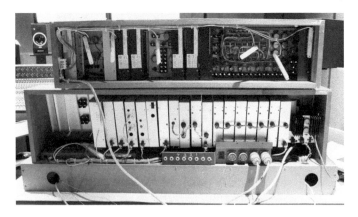

Figure 3.4 View of the Back

Figure 3.5 Moog Keyboard

During early development, the topic of how to trigger the synthesizer came up frequently. Early iterations utilized unorthodox methods such as push buttons and ribbon strips. The idea of using a piano-like keyboard was floated but was met with hesitance. In fact, early electronic music composer Vladimir Ussachevsky, who was working with Moog and Deutsch, whole-heartedly disagreed with using a piano style keyboard. Ussachevsky felt that the inclusion of a piano-style keyboard would constrain musicians into treating the synthesizer as a traditional western instrument and play piano inspired passages on it. Ussachevsky felt that this went against the purpose of creating a new instrument that was not bound by traditional western music norms. In the end, the piano style keyboard won out and this proved to be one of the defining moments in the synthesizer's future of being used as both an orchestra emulator and an orchestral instrument in and of itself.

Switched on Bach

In the early days of synthesis, these instruments were viewed more as avant-garde sci-fi noise boxes rather than musical instruments. Although the potential to create both authentic and new sounds was great, many musicians felt that both the learning curve and price of these synthesizers was far too high to justify purchasing what was often considered a glorified noise maker. This sentiment however, quickly changed in 1968 with the release of arguably the most influential electronic record of all time – *Switched on Bach*.

In 1967 Wendy Carlos, a relatively unknown composer, got the idea to record a number of Bach songs solely on a Moog. Carlos, who was very interested in the promise of electronic musical instruments, felt that the current cannon was lacking appealing music that could be enjoyed by the masses as most synthesizer music at the time was extremely *avant garde* and experimental. The recording was not only instrumental in the advancement of the synthesizer as an instrument but also proved to advance recording technology as a whole due to the complex nature of the recording sessions.

Figure 3.6 Close-Up

Figure 3.7 Moog Model 15 App

Carlos worked closely with Robert Moog during the recording process. Moog was constantly taking in Carlos's suggestions and tinkering with and modifying the system to better suit her needs. Recording proved to be extremely tedious. Because the synthesizer was monophonic (capable of playing only one note at a time), Carlos had to record each note of a chord or harmony separately and layer them – a feat in and of itself when recording on an eight-track tape machine! Besides the tedious recording process, the Moog needed to be constantly checked for drifting (when an oscillator changes pitch and becomes slightly sharp or flat).

In the end, Carlos recorded ten pieces over the span of five months and released the album in the fall of 1968. The album proved to be a success. By 1969, *Switched on Bach* was on the top 40 charts and by 1974, sold over 1 million copies. The album was also a success with industry professionals. Shortly before the album's release, Moog spoke and previewed one of the songs for the annual Audio Engineering Society conference where it received a standing ovation.

Moog saw a huge jump in sales of their synthesizers immediately after the success of *Switched on Bach* as studios and composers across the county began utilizing these remarkable machines. For the first time since their inception, the synthesizer was starting to be treated as an actual, useful instrument. It was also during this time that synthesizers were beginning to be utilized as an orchestral instrument. The legendary film composer John Barry utilized a Moog modular and featured it heavily on the score of the 1969 James Bond classic *On Her Majesty's Secret Service*. Although synthesizers were featured in many science fiction film scores previously, Barry was one of the first to use the Moog as a melodic instrument in a film score.

As the decade ended, the synthesizer was finally being viewed as a useful melodic instrument and more and more mainstream artists and composers were beginning to utilize the instrument in their work. As more Moog systems were being sold, other companies began cashing in on the emerging synthesizer craze. Roland, Arp, and E-mu soon began offering modular systems of their own. The desire for synthesizers was reaching an all-time high but the cost, size, and technical know-how was still too

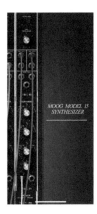

Figure 3.8 Model on the iPhone

prohibitive for most. Moog once again was at the forefront of answering this call and quickly came out with arguably the most famous synthesizer of all time – the Minimoog Model D.

The Minimoog finally brought synthesis to the masses. It was relatively compact, relatively affordable, and most importantly did not require a huge amount of synthesis knowledge in order to operate. Unlike its modular predecessor, the Minimoog was "normalled" meaning it didn't require patch cables to create sound – everything was pre-wired inside the case. The Minimoog was hugely successful. It seemed as if every studio, and band had at least one Mini. Again, other manufacturers were not far behind with Arp creating the 2600 and Odyssey synthesizers, Korg with the MS-20, Oberheim with the two-voice, Octave with the Cat, EDP with the Wasp, and so on and so forth. For the first time, synthesizers were easily accessible and being heavily used.

Additional instruments

As synthesizers were becoming more available, a man named Malcom Cecil was starting to fill a studio with various synths and interconnecting them. Eventually, he created what he called T.O.N.T.O. (The Original New Timbral Orchestra). T.O.N.T.O is a multitimbral polyphonic synthesizer made up of various Moog Modulars, Arp 2600 s, Oberheim SEMs, and various other synthesizer components all wired to-gether in order to be played by one or two engineers. Cecil released a couple of albums fully recorded on T.O.N.T.O. but also featured T.O.N.T.O on famous artist's albums such as Stevie Wonder. By having so many unique voices, Cecil was able to program full pieces of music and have various sections of the system designated for percussion sounds and other sections for melodic and droning sounds. T.O.N.T.O. helped pave the way for modern electronic music DAWs and work-stations that feature various instrument types that can be recalled at any time by the user.

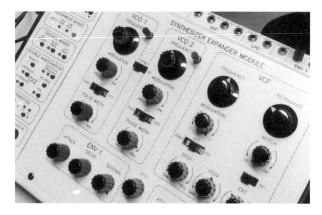

Figure 3.9 Oberheim Synth

As the 1970s wore on and music tastes changed, the desire for orchestral instrumentation in popular music was growing. There were bands such as Chicago and Jethro Tull that featured various orchestral, "non-traditional" pop instruments such as brass sections and flutes, but the vast majority of artists would have to settle for studio musicians who could play these instruments during recording sessions. It was not long before artists began programming brass or string like sounds on their synths. Synthesizer manufacturers were quick to realize the commercial potential of synths that were designed to emulate orchestral sounds. Manufacturers like Moog and Arp began producing these types of synths, most often referred to as string machines due to the vast tendency for them to be orchestral string based. These string machines acted as a less expensive, less finicky, and overall more versatile replacement for the aging Mellotrons and they became extremely popular among both studios and musicians. These synths can be heard in most pop music of the era but especially in the big disco hits of the late 70 s and early 80 s.

Although these string synthesizers and others like them satisfied a need for orchestral sounds without having to hire an actual orchestra, no one could honestly claim that they sounded like real strings. They often sounded like cheesy synthetic interpretations of

Figure 3.10 Logic's ES E – String Synth

Figure 3.11 DX7

strings. Although this sound was unique and cool in its own right, musicians and engineers were still striving for more realistic sounding orchestral instruments.

FM synthesis

The next big advancement in synthesis technology was the introduction of FM, or frequency modulation, synthesis. FM synthesis creates its sounds via modulating a waveform, known as a carrier, with a modulator signal. FM synthesis is capable of creating either harmonic or non-harmonic sounds and can create some of the most complex sounds available in synthesis. Due to the unstable nature of analog circuitry, FM synthesizers typically employed digital circuitry making them the first digital synthesizers. FM synthesis was originally only seen in universities due its expensive and complex nature but that all changed in 1983 when Yamaha released the hugely popular DX7.

The DX7 was arguably the first widely popular digital FM synthesizer. One revolutionary feature of the DX7 was the vast amount of factory preset sounds that came loaded on it. The combination of numerous presets without a crowded face was a welcome change for many musicians fed up with the finicky tweaking and numerous knobs of traditional analog subtractive synthesizers. For the first time, musicians could simply press a button to recall a totally new sound without needing any synthesis knowledge whatsoever. Ironically, most people ended up using the DX7 for its presets and ignored the extremely powerful FM synthesis engine that was capable of truly amazing sound design. The DX7 helped satisfy the desire for orchestral sounds as it shipped from the factory with a number of orchestral presets including three brass sounds, three string sounds, a full orchestra patch, marimbas, chimes, timpani, and a flute just to name a few. These sounds showed up on a huge number of popular tracks and film scores of the time cementing the DX7s legacy as one of the most popular synthesizers ever produced.

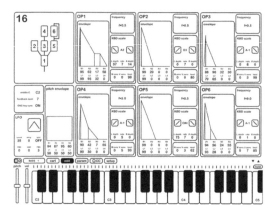

Figure 3.12 KQ Dixie – DX7 Clone for iPad

Perhaps it was the prevalence (some would say overuse) of the DX7 in popular music, or perhaps it was the daunting task of learning how to edit patches on the DX7, but either way musicians inevitably grew bored of the orchestral patches and continued to crave more realistic orchestral sounds. Their call would soon be answered in the form of a revolutionary new technology – digital, sample-based synthesizer workstations.

Early digital workstations

At this point in audio technology history, analog tape was still in its prime and computers were still in their infancy. However, two companies, Fairlight and New England Digital, were introducing systems: the Fairlight CMI and Synclavier respectively. Both of these machines utilized new digital micro-processors which enabled for real-time digital re-cording, manipulation, and reproduction. These two systems eventually paved the way for digital audio workstations (DAWS) as we currently know them.

The Synclavier and the Fairlight were both capable of digital sampling as well as complex Digital synthesis. Both machines allowed users to either create new sounds with their complex FM or additive synthesis engines. These sounds could be stored on a variety of different mediums such as hard and floppy disks. These stored sounds could then be manipulated further through various "sweetening" functions on the systems. However, where these machines really shined, were their ability to digitally record, and manipulate acoustic sounds. Users could record any variety of real-world sound and load

Figure 3.13 Fairlight CMI iPad Clone

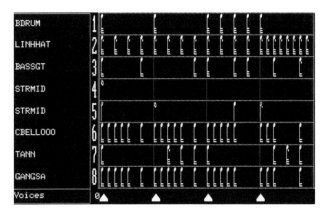

Figure 3.14 CMI Sequencer Clone

it in as voice on these systems. From there, the sound could be mapped across the entire key bed, raising in pitch as the user played higher and lowering in pitch as the user played lower. Although this is what any cheap, modern, sampler can do, the feat was revolutionary at the time. However, both systems allowed for much more creativity and flexibility than simply sampling sounds. Once the user loaded up the desired sound, they could digitally manipulate the sampled sounds by cutting out sections, looping sections or even re-drawing waveforms. These sounds could then be stored and programmed into one of the overly complex on-board sequencers for recording and performance.

For the first time, musicians and engineers were finally beginning to feel like they had realistic instrument sounds at their fingertips. These machines however, were not cheap. Both the Fairlight and Synclavier were completely cost prohibitive for all but the most successful musicians and largest studios. Despite their large price tag though, there were a number of famous users such as Frank Zappa, Herbie Hancock and legendary sound designer Ben Burtt just to name a few. In fact, Frank Zappa produced a number of albums later in his career solely using his Synclavier for all instrument parts.

The sounds both the Synclavier and Fairlight were capable of reproducing were so faithful to their respective instruments that it caused a number of musicians to place disclaimers on their albums stating that the music enclosed was played by actual musicians using actual instruments. Although extremely realistic orchestral sounds were finally available to musicians, the steep learning curve and price tag still kept this technology out of the hands of most musicians. There was a void in the market for an affordable digital sampler and a number of companies began producing just that. However, the company E-mu jumped to the forefront with their legendary Emulator models.

Figure 3.15 EMU Units

Emulators

The very first Emulator was a very basic sampler that allowed musicians to sample audio and store it on floppy disks. Unlike the Fairlight and Synclavier however, the Emulator did not offer the user much in terms of editing and customizability. The second Emulator, known as Emulator II was released in the mid-1980s to very high acclaim. By the release of the second Emulator, there were a multitude of highly respected third-party sample libraries that could be purchased which made the Emulator that much more desirable. Besides the large number of samples available to users, the Emulator II featured a number of analog resonant filters which allowed the user much more flexibility than its predecessor when shaping sounds. The overall fidelity of the sampler was also much better than its predecessor which when placed in combination with the editing power and large sound libraries available, made the Emulator II a huge success.

Piggy-backing of the success of the Emulator systems, many other companies started producing cheaper more affordable samplers. Companies such as Casio, Akai, and Ensoniq were producing samplers for less than $2,000. Although sound fidelity and customizability arguably suffered in these cheaper samplers, the low-price points more than made up for it.

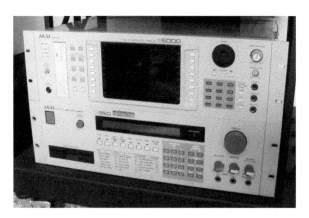

Figure 3.16 Akai Samplers

Throughout the 1980s, orchestral instruments heard on the majority of pop records were created using one of the systems listed above. Whether using the DX7, the Fairlight or Synclavier, or the more affordable Emulators, sampled or synthesized orchestral instruments were everywhere. Musicians no longer felt trapped by the instrumentation in their group. If a bagpipe, zither, violin, sitar, timpani, or cello was wanted on a song, all the artist had to do was recall the sound. This era of creative flexibility served to almost rejuvenate popular music and breathe fresh air into music that had largely been made up of the same instruments since the advent of recorded sound.

Improving samplers

The advent of powerful affordable samplers also served to change the way music was written for movies. Composers were able to play their scores for directors as they would be heard from a full orchestra instead of just playing the main themes on a piano as had been tradition. Directors were now able to incorporate the score more deeply into movies by allowing various themes to dictate artistic choices during filming and editing. With the advent of this technology, music and film solidified the symbiotic relationship they share today.

In addition to making the scoring process more collaborative, these samplers also allowed smaller budget films to have more powerful and expensive sounding scores. Filmmakers like John Carpenter were able to create their own scores for their films, completely bypassing the budget typically reserved for the composer and scoring sessions with the orchestra.

For the next decade, samplers continued to become more affordable while increasing in storage capacity and usefulness. Many companies were offering standalone samplers as well as samplers incorporated into digital hardware workstations. The advent and rapid implementation of MIDI brought with it keyboard controllers and sound modules which allowed for a huge amount of "from-the-factory" sounds. Artists and engineers were finally able to faithfully reproduce any instrument in the studio or in a live setting.

Figure 3.17 EXS-24

Digital audio workstations

The quality of samples and synthesis driven sound reproductions have increased greatly with the advent of DAWs and personal computers. Programs like Logic, Ableton Live, Pro Tools, and FL Studio all come with a huge bank of orchestral sounds that can be recalled instantly. Unlike many old samplers, these sounds can be heavily manipulated and altered to the artists' taste. In the hands of a skilled engineer, the average listener would be hard pressed to tell the difference between a recording made with a physical orchestra versus one made entirely digitally.

Although sampling is typically thought of as the best method for recreating orchestral sounds, many modern synthesis engines can be used to successfully mimic orchestral instruments. Dating back to the earliest days of synthesis and the earliest string machines, subtractive synthesis was the first synthesis method used to create orchestral sounds. Although many people grew tired of the relatively basic sounds created on early machines, subtractive synthesis still remains an extremely powerful synthesis engine for sound re-creation especially when paired with more modern control methods such as polyphonic after touch, breath control, and controller assignability.

Subtractive synthesis

At its essence, subtractive synthesis works by taking an extremely rich harmonic signal and taking out (subtracting) content in both the harmonic and amplitude domains. When designing sound using subtractive synthesis, you start with an oscillator. An oscillator is a sound producing circuit and can be thought of as the heart of a synthesizer. An oscillator produces a waveform that can be altered later. Although all oscillators are different, most will be capable of producing sine, triangle, sawtooth, and square waves. These waves vary in their harmonic content and all sound different.

Figure 3.18 Kawai SX-240

Figure 3.19 Basic Waveforms

Waveforms

Let us take a moment to explore the various waveforms. Becoming familiar with each of the waveforms and how they sound by themselves, as well as how they react to different settings of the mixer, amp and filter, is key to creating realistic patches. In order to understand why these waveforms sound different from one another, we must first examine timbre and the harmonic series.

Timbre can best be described as the unique character of a sound – a piano sounds different from a trumpet playing the same note because of each instrument's timbre. An instrument's timbre is caused by a variety of factors such as the material of the instrument, sympathetic vibrations, overtones, and the way it is played. For the purposes of this book, we will take a closer look at overtones.

Overtones

Overtones, put simply, are frequencies above the fundamental frequency being produced when an instrument is played. For example, concert A on a piano is 440 Hz. However, 440 Hz is not the only frequency we hear; we hear many higher frequencies (overtones) at the same time. The amplitude of these overtones is key in making up the instrument's timbre. In music, these overtones are rarely random. In fact, these overtones typically follow a very strict pattern known as the harmonic series.

The overtones in the harmonic series are all integer multiples of the fundamental frequency. In the harmonic series, the fundamental frequency is considered the first harmonic whereas the second harmonic is exactly two times the fundamental, the third harmonic is three times the fundamental and so on and so forth. The varying amplitudes of these upper harmonics play a large role in the overall timbre of the sound.

First Harmonic (Fundamental)

Second Harmonic

Third Harmonic

Fourth Harmonic

Fifth Harmonic

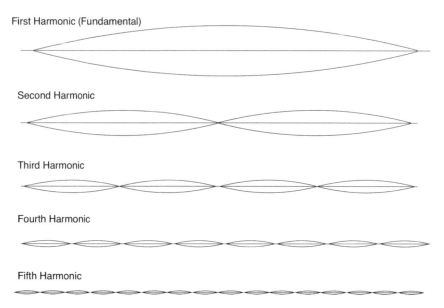

Figure 3.20 Harmonic Series

The various waveforms an oscillator produces are considered simple waves in that they produce harmonics in a predictable manner both in frequency and amplitude. A sine wave is the most basic of the waveforms. A sine wave is just the fundamental frequency; it contains no upper harmonics. Next to the sine wave, the triangle wave is the next most basic sounding waveform. A triangle wave features only odd harmonics (3,5,7,9, etc.). The amplitude of these harmonics drops at a rate that is proportionate to the inverse square of the harmonic number. Therefore, the third harmonic is $1/6^{th}$ the amplitude of the fundamental, the fifth is $1/10^{th}$ and the seventh is $1/14^{th}$ the amplitude of the fundamental. This causes very few harmonics to be audible resulting in a waveform that is fairly close in sound to the sine wave.

The next two waveforms are significantly more harmonically rich. A square wave, like a triangle wave, only features odd harmonics. Unlike a triangle wave however, the square wave's harmonic's amplitudes are just inversely proportionate to the harmonic number. This means the third harmonic is $1/3^{rd}$ the amplitude, the fifth harmonic is $1/5^{th}$ the amplitude, and the seventh harmonic is $1/7^{th}$ the amplitude. Because the harmonics are much louder in relation to the triangle wave's harmonics, the square wave is much more harmonically rich. Sawtooth waves are the most harmonically rich of the basic waveforms. Sawtooth waves feature all harmonics (both even and odd). The harmonics of a sawtooth wave follow the same amplitude drop as the square wave (i.e., the second harmonic is half the amplitude, third harmonic is $1/3^{rd}$ the amplitude, fourth harmonic is $1/4^{th}$ the amplitude etc.)

Figure 3.21 Additional Waveform Example

Recreating sounds

It is typically good practice when re-creating a sound to start with a waveform that sounds somewhat like the sound being re-created. Often times, synthesizers will feature multiple oscillators which can be combined at various octaves and waveforms in order to create more complex starting sounds. The output of the oscillators is then routed through a mixer where their relative amplitudes can be controlled.

From the mixer, the sound is then routed into a filter – typically, a resonant low pass filter. Upper harmonics can be taken out of the sound causing it to sound darker as the filter is rolled off. Some animation and character can be added to the sound through the resonance circuit of the filter which feeds the cutoff frequency back into the filter causing an emphasis at that specific frequency.

Figure 3.22 Oscillator Mixer

Figure 3.23 Filter

The next thing that must be adjusted is the synthesizer's loudness envelope. When a physical instrument is played, sound does not instantaneously start and end. Instead, the sound ramps up to its loudest points and then dies off at rates dependent on how the instrument is played. This is known as a sound's envelope. The same effect can be achieved on a subtractive synthesizer using an envelope generator. This device is a circuit which allows you to control multiple stages of a sound's loudness such as attack, decay, sustain, and release. In this scenario, the envelope generator is routed to the synthesizer's amplifier in order to control the loudness of the synth.

Although an envelope generator is always going to be routed to an amplifier, many subtractive synthesizers feature multiple envelope generators allowing for higher levels of customizability. For example, routing an envelope generator to the synthesizer's filter would result in the filter opening and closing in relation to the settings of the envelope generator. Routing an envelope generator to the pitch control on an oscillator on the other hand, would result in the pitch changing in relation to the shape of the ADSR.

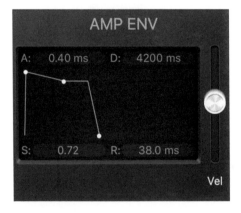

Figure 3.24 ADSR - Envelope

Figure 3.25 Low Frequency Oscillator

Another point of customizability on most subtractive synthesizers is the ability to modulate various parameters through an LFO. An LFO, or low frequency oscillator, is an oscillator which produces waveforms at a lower pitch than humans can hear. These waveforms can then be routed to the control inputs of various parameters causing a repetitive modulation. A simple vibrato can be created in this way by routing a sine or triangle wave LFO to the pitch control of the oscillators at a low setting. These LFOs can typically be routed to a variety of locations on the synthesizer including the amplifier and filter resulting in a large amount of sonic potential.pt

Cello example

Let us explore how we can use subtractive synthesis for sound re-creation by trying to mimic the sound of a cello. First, we must analyze a cello. We must ask ourselves a series of questions: is a cello rich in harmonics or is it basic? How does the sound get set into motion (i.e. plucked, bowed)? How does the amplitude change over time? Does

Figure 3.26 Complex LFO Setup with Modulator

Cello Waveform   Cello Waveform Zoomed In

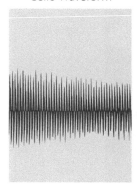 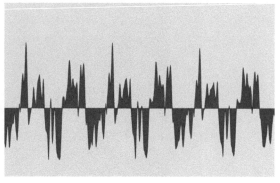

Figure 3.27 Cello Analysis

the pitch stay constant or does it waiver? Besides listening to the sound and asking yourself these questions it is often helpful if you can see a spectrum analysis of the sound.

Looking the spectral analysis of the cello recording, it is evident that the sound has a very present low frequency. Therefore, we will begin by placing our first oscillator in a slightly lower octave – such as 16'. Looking at the harmonic content of the sound, it is also evident that the cello is fairly rich in harmonics. Therefore, let us set the oscillator to a saw tooth wave – the most harmonically rich of the waveforms. The raw saw tooth sound is much too harsh. We must take off some high frequency with the low pass filter. How much to take off will depend on the synthesizer itself as well as the cello you are trying to imitate.

Next let us listen to how the sound changes in amplitude overtime. Typically, a cello string will be set into vibration via a bow being scraped across the string. Therefore, it takes some time for the sound to get to its maximum amplitude. We can imitate that by slowing down the attack portion of the amplifier's envelope generator. After being set into vibration by the bow, the sound will ring out once the player stops playing. The sound will quickly dissipate and fade out through the body of the cello. This same thing can be achieved by setting the decay, sustain, and most specifically release time of the envelope generator.

Once we are satisfied with our envelope generator sounds, we can examine the core oscillator sound itself. The cello has many higher order harmonics at play. Therefore, it might be beneficial to add a second oscillator at a higher octave – say 8'. This second oscillator shouldn't be overpowering as its acting as an upper harmonic source to the already rich sound from the saw tooth wave. Therefore, setting the second oscillator to a triangle wave and turning down its level in the mixer might be beneficial.

This should be getting us close to a cello sound. However, there are still a few more things we can do to improve the realistic nature of the sound. One of these things is pitch modulation. When examining a cellist, they often times add vibrato to their playing by slightly moving their finger from side to side on the finger board. Although this is simply a performance technique and is not always used, mimicking this vibrato can greatly improve our synthesized cello sound by making it sound a little less robotic. As stated earlier, we can achieve pitch modulation by routing an LFO to our oscillator. Since we are using two oscillators for this sound, we will have to make sure that the LFO is being routed to both oscillators simultaneously. Typically, a sine or triangle wave shape is utilized on the LFO since they are both the "smoother" sounding waveforms for modulation as they go up and then back down at equal rates. Since we only want to use this modulation as a form of slight vibrato, we need to make sure that the modulation amount is set fairly low. This would be done on either a knob or modulation wheel depending on the synthesizer.

Next, we can examine how the cellist moves from note to note. Often times, their finger will glide up or down the fingerboard allowing us to hear slight variations in the note as the slide up or down to it. We can achieve a similar effect on most subtractive synthesizers by employing glide (sometimes referred to as portamento or glissando). Glide allows one note to slide into another note when two different keys are pressed. This effect gets more drastic as the spacing between the notes increases. Adding a tiny amount of glide can help make the patch sound a little more organic.

Lastly, we will add reverb to our patch. We are accustomed to hearing orchestral instruments such as cellos in concert halls where the rooms are often times easily excited and acoustically beautiful. Therefore, it is only natural to try and impart this characteristic onto our patch. Some synthesizers will feature reverb on the unit itself or it must be added as an effect but reverb can go a long way in making our sound much more believable. Choosing a nice sounding reverb is crucial. I personally like convulsion reverbs which utilize impulse responses taken from physical spaces to model an artificial reverb.

With these settings, we should be getting fairly close to an accurate sounding cello. However much of the believability of the sound will rely on the performer themselves and how they utilize phrasing and performance tools such as volume wheels/pedals and the like. One thing that should go without saying but often times eludes even the best of programmers is making sure to stay in the given instrument's range. For example, a typical cello is only capable of playing from C_2 to A_5. Therefore, when playing or programming a cello patch on a synth, it is important to stay in that range. In the same vein, it is also important to play passages that a cellist would be able to play meaning its best to refrain from huge jumps up or down the fingerboard as well as chords.

Flute example

Let us now attempt to mimic a flute. First, we must analyze a flute. Since we are dealing with a flute, we can assume our sound will be in a higher register (this can be verified by a quick look at a frequency analysis of a flute recording).

Flute Waveform   Flute Waveform Zoomed In

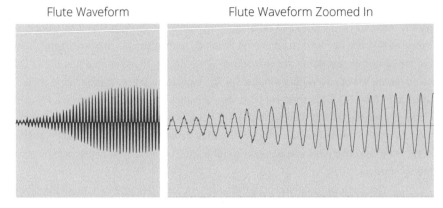

Figure 3.28 Flute Analysis

It is apparent that the flute is not nearly as harmonically rich as the cello from the previous example. Therefore, we will set our oscillator to a triangle wave. The flute has a similar attack time to the cello due to the sound being initiated via breath so we must slow the attack time down. Next, we will lengthen the release. Since the flute is played with breath, we can add a little noise to the sound in order to mimic this and add some realism. Most synthesizers will feature a noise generator somewhere near the oscillator section with both white and pink noises being common options. We'll turn the noise fairly quiet in relation to the main oscillator sound.

In a similar fashion to the cello, adding a slight amount of pitch modulation to our sound can help make it sound less stagnant and robotic. In addition to the pitch modulation, we could alternatively add a slight amount of amplitude modulation, or tremolo. This is accomplished in the same way as the vibrato. In this case however, an LFO is routed to modulate the amplifier so that the volume goes up and down in correlation to the wave shape of the LFO. This can help mimic the effect varying breath strength has on the flute.

This same technique can be utilized for recreating virtually any orchestral sound. The more complex the sound however, the more complex the synthesizer must be. When using subtractive synthesizers for complex sound recreation, large modular synthesizers are typically better. Large modular systems will typically have many more oscillators, envelopes, and amplifiers than standard synths. This allows for more in-depth recreation where various harmonics and their individual envelopes can be matched with much greater accuracy.

Whether using a physical synthesizer or a virtual instrument, subtractive synthesis is by and large the most utilized synthesis format for sound recreation. This is due in large part to the user-friendly interface of most subtractive synths as well as the tactile nature of hearing the change to the sound as the user adjusts parameters. However, really any synthesis format can be used to recreate acoustic and orchestral sounds.

Figure 3.29 Sculpture Instrument

Physical modeling

Physical modeling synthesis is one of the more powerful synthesis types for mimicking acoustic instruments. Physical modeling generates waveforms using a number of complex algorithms which are used to compute mathematical models of physical sounds. These models are then used to simulate the sound of the instrument. Due to the complex nature of the synthesis format itself, however, it is not as widely utilized when creating sounds from scratch but rather used more heavily on preset driven sounds.

Physical modeling can offer users more versatility in an emulation synth. This is particularly true in the velocity domain. With sample-based synthesis, the user is stuck with however many varying velocities were recorded of the instrument. Using an acoustic piano as an example, the sampling engineer would record individual piano notes at varying intensities and program them to various velocity ranges on the VST. This leaves the end user with a limited range of expression when playing the instrument. A physical modeling synthesizer on the other hand can apply a varying degree of intensity to every single resolution point of velocity available on the MIDI standard. This offers the end user a huge amount of expression. Another area in which physical modeling synthesis shines is mimicking the way in which a particular sound is set into motion. All instruments are played utilizing different means. Some instruments use bows, some mallets, some plectrums. A subtractive synthesizer is limited to standard ADSR envelope generators when trying to mimic this quality. A physical modeling synthesizer on the hand uses the same mathematical models mentioned earlier to faithfully reproduce the unique characteristics of different sounds being set into motion and excited.

Although not as prevalent as subtractive synthesis, there are some extremely popular PM synths on the market. Although some hardware physical modeling synthesizers have been produced, most PM synths exist in the virtual domain. In fact, many DAWs either come with physical modeling synthesizers or have ones that can be purchased. Examples of these DAW based physical modeling synthesizers are Logic's "Sculpture," Image Line's "Sakura," and Ableton's "Collision." In addition to these DAW based PM synths, many third-party companies such as Applied Acoustic Systems (AAS), Antares, and Xhun, offer physical modeling synths that can be purchased and used with most popular DAWs.

Figure 3.30 Sakura Instrument

Whether using a synthesis engine or using samples to create orchestral sounds, modern DAWs and synthesizers give the modern musician a huge pallet to work with when composing and performing music. Although some people may think that easy access to orchestral sounds will eventually bring demise to symphonic music played by humans, I believe the opposite to be true. It is by giving musicians easy access to these sounds that will keep symphonic music alive; or at least contribute to its continued success. By having access to realistic sounding orchestral and real-world instruments, composers can feel free to write music for orchestral instruments when they might not have, were it not available to them. By writing pieces specifically designed for orchestral instruments, orchestras the world over can play this music even if the source material was recorded using one of the aforementioned synthesis techniques. One just has to look at the success many local orchestras have playing the cinematic music of John Williams or Hans Zimmer. Easy access to these sounds will, in my opinion, keep the orchestra alive for many more years to come.

Additional examples

In order to fully understand the potential of synthesis in the orchestration process here are three examples which incorporate the principles listed earlier in this chapter.

Modeled instrument

There will certainly be a day when a fully modeled instrument might be able to be indistinguishable from its acoustic counterpart, but it has not happened yet. In the meantime, modeled instruments are still a great way to add layers to other instrument sections, either as a similar sounding patch or something completely unexpected. In this example Logic Pro's Sculpture is used to add a string-like sound to a string section. Sculpture uses a modeled "string" as the central part of its sound engine, which is only loosely a string in any sense of the word. It can be made of a range of materials, activated by a range of

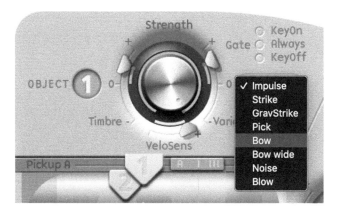

Figure 3.31 Object 1

objects and actions, and then processed through a deep set of effects such a modeled instrument bodies, filters, and other comprehensive tools. Mastering this instrument is a massive undertaking but even using some of the presets can be useful in music production.

The first step is to assign the bow object to the string. This mimics a stringed instrument's bow and has a representative sound. Next, set the string material to a mix between steel and nylon. Setting it all of the way to full steel has an overly metallic sound, which is tempered by adding in some of the nylon material. After setting the material then the density and resonance are adjusted according to personal taste and the desired end sound. In the end, adjusting the material so that it shifts in pitch a little bit at the start of the attack. The release is set so that there is a short amount of resonance when the key is released. After that the body is assigned, which doesn't necessarily have to be the obvious choice… try a few of them out and decide which one adds the desired sound.

The last phase of this process is to tweak the parameters until the sound works. That part usually comes after the combination of traditional instruments and the Sculpture instrument takes place, since this is when the needs of the sound is clearer. If the attack of the sound should blend in more, then the string material can be softened,

Figure 3.32 Materials and the Body

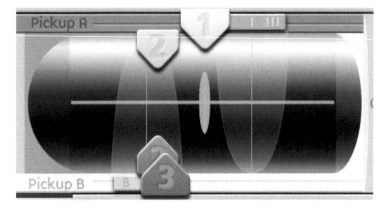

Figure 3.33 Adjustable Pickups

or the power of the original bow object can be adjusted. There are also "pickups" on the string which are similar to guitar pickups, and these can be moved to change the sound.

When paired with a string instrument or section Sculpture can be easily adapted to match the same contour of the instruments. One benefit of working with MIDI is that it can be copied from one track to another for easy doubling. If Sculpture starts later than the strings, then adjust the attack so it starts faster. Adjust the overall level so that Sculpture blends in nicely and adds to the texture without overpowering the string sound.

There is a plethora of options when using Sculpture to blend with real instruments. Objects can be set to blow like flutes, drop like percussion, or pluck like pizzicato. There are a ton of possibilities. On top of all of this, Sculpture can do something which very few string instruments can do – it can morph mid-note. Using a five-point morph matrix, which stores five different instrument settings and as the control point is moved around the square, the sound morphs between the settings. It can even be programmed to consistently start at one point and then move in a predetermined path around the morph pad. This could be useful when motion is desired in an instrument section to add brightness, intensity, or other timbres over time.

Figure 3.34 Morph Pad

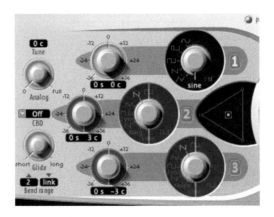

Figure 3.35 Sine Wave

Jumping ahead to the topics in Chapter 4, the morph pad can also be controlled using various MIDI controllers, including one where it involves waving hands in the air. With all of the amazing technology available, the only thing to be cautious about is that the tech doesn't distract from the music making process. If the technology is able to enhance the process then use it as much as desired. Is it worth facing the learning curve of an instrument such as Sculpture? Certainly, there are features which exceed what the average composer might need, but the more that is known about how it works then the more flexible and powerful the instrument can be. The magic of synthesis is most apparent when the skill of the programmer is equal to that of their imagination. In other words, if a person can think of a sound and then create it using their tools then it is clear that mastery of the process exists. Until that happens then presets suffice to get the instrument as close as possible without having the skill needed for true customization.

Subtractive synthesis

There are some simple but effective uses for a subtractive type synthesizer in the orchestration process. Perhaps the most important use is for the sine wave wave-shape, which is a pure tone that can be reproduced at a low frequency and added into the music as an effect indicating a foreboding emotion or representing certain types of intense action. The sine can be tuned to be an extension of the harmony of the music as a fundamental root.

This example could be accomplished with most synths but the ES2 in Logic Pro is used as the instrument. The ES2 synth has three oscillator sources, which can be mixed between in the triangle shaped source mix area. Only one is needed, set to sine wave. This wave-shape is the simplest of all wave shapes with no harmonics and therefore most of the other effects on the synth do not make much of difference to the final sound since most effect rely on harmonic content to be effective.

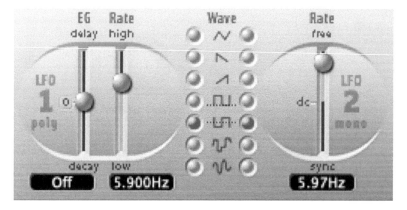

Figure 3.36 LFO Flutter

In the orchestration, either clone the lowest instrument part and use that to trigger the synth or add a drone part which stays on a common low tone throughout a section. The trick is to use this sparingly so that it is impactful when it does occur. There are other effects which a subtractive synth is useful for creating, such as rhythmic patterns and other sound effects. The sound can be modulated using its built in LFO or by an outside modulator. For instance, the output volume could be turned on and off in rapid succession in sync with the tempo. The resulting flutter adds an interesting texture to the other textures of the orchestra.

FM synthesis example

During the 1990s, FM synthesis was highly successful because it was one of the few types to be able to replicate real instruments in a semi-realistic manner. Looking back now it is clear that they were not actually that close to reality but the sound of the DX7 was users on countless pop/rock albums and is still widely recognized. For this example, the default bass and electronic piano patches from the DX7 are used to add some nostalgia to the sound of the project. These sounds might be layered in or used independently.

Figure 3.37 KQ Dixie Connected via IDAM+

Figure 3.38 Retro Synth in Table Mode

While it is possible to obtain a DX7 synth online for only a couple hundred dollars, it is likely prohibitive to buy it just for a few sounds. There are several software versions of the DX7, or of similar sounds, which makes it easy to obtain for regular use. One affordable version is a cone of the DX7 which runs on iOS and is a perfect match. The section in Chapter 4 covering iDAM+ explains how to use an iPhone or iPad in the production process. Another use of FM synthesis is in creating metallic sounds and bell-like sounds. These are great for creating ethereal parts in the music, with a special focus on film music.

Wave table synthesis

The final two types of synthesis showcased here in the example section are perhaps two of the most applicable to orchestration because of their ability to use source material which is from the orchestra itself. A wave table is an audio file which is compilation of smaller sounds or just a longer sound broken into smaller sections, which is then used as the source engine for synthesis techniques.

This type of synthesis is very useful in creating industrial sounds and heavy percussion, but while those are useful there is a different technique of interest in this situation. By loading small excerpts of instrument recordings as wave tables the entire synthesis process has the potential to sound like the original sample, be ripped apart into something completely new, or bent into some combination of everything involved. When a file is imported as a wave table, the key process is in playing a single element of the wave by repeating it over and over. The instrument can also play through the whole set of waves, reproducing the original sound. It can be played forward, backward, at different speeds, or in various patterns. The wave tracking element is flexible enough to sound pretty or pretty intense.

Figure 3.39 Importing a Table Options

Figure 3.40 Nave Instrument

This example involves a short second-long recording of a cello note. After trimming the source, it is imported into the Vintage Synth in Logic Pro. There are different analysis types for the import, but the default is acceptable. Because the sound engine breaks the sound apart it is able to relatively easily pitch-shift across the range and it takes a single file and turns it into an entire instrument. The next step is to adjust the tracking speed so that the instrument sounds closer to the original file or turn it into something different. One powerful way to utilize the tracking function is to attach the speed to an external controller so that at a resting position the sound mimics the original file and is closer to a cello, and then the speed can be increased to warp the sound. This layers well with a string section but can quickly morph into something else. A really powerful iOS wave table synth called "Nave" should be in the arsenal of every mobile music maker.

Re-synthesis

In a very similar process to the previous example, re-synthesis takes a single file and creates an entirely new version based on the original. It is possible to do this with a few types of synthesis, but this example uses additive synthesis as the primary engine. "Alchemy" is one of the most powerful sound tools ever made and it is capable of incredible things. In this example Alchemy is used to import a brass sound, synthesize it into a new sound, and allow complex performance options so it can be used in the production process. Perhaps the term re-synthesize is either redundant or not exactly accurate, but since "synthesize" is too general under the circumstance of discussing the broader topic of synthesizers, re-synthesis seems to be appropriate.

Additive synthesis in this case analyses the imported audio sample of a

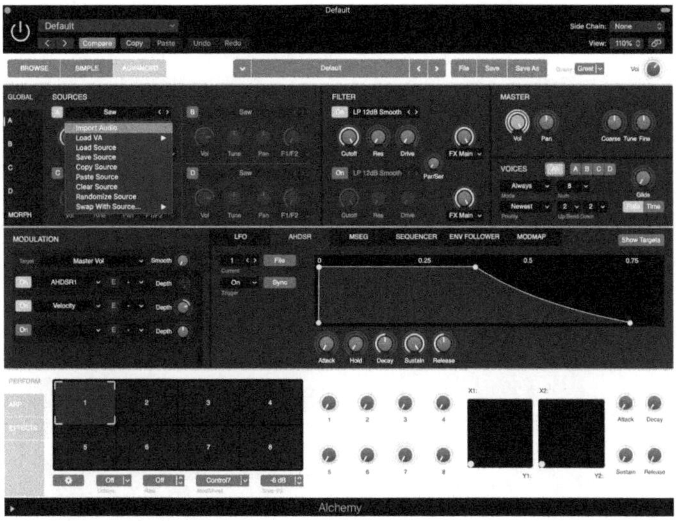

Figure 3.41 Alchemy Instrument

trumpet and uses a series of tones to create the new version. It is possible for this to sound related to the original recording, but it also tends to vary from analysis to analysis. The important feature is that once the additive process takes place then the instrument has created a full version across the entire keyboard. Traditional sampling creates instruments which pitch shift samples, but the resulting files are lengthened or shortened to accomplish it. This re-synthesis results in "pitch shifting" which takes a single file that stretches across the entire range without changing the timing on any note. It is a compromise no matter which tools are used but re-synthesis is among the most useful due to the ability for it to create something new, but something which can be molded

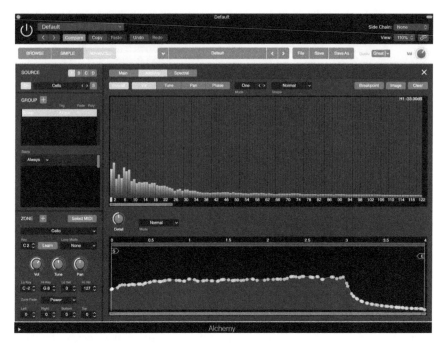

Figure 3.42 Additive Synthesis Edit Window

and adapted into so many different sounds. The edit window showcases additional manipulation features that come with drawing tools and other editing tools.

Summary

Synthesis is a powerful music creation tool and one which is able to work side-by-side with other orchestration instruments. It can make the final mix sound larger than life, other worldly, or simply cause the listener to feel nostalgic for music of previous decades.

Chapter 4

Performance tools

Performance tools are a critical part of the digital orchestration process. While it is possible to create an entire digital orchestration with a mouse, it becomes a question of efficiency and musicality, both of which are enhanced when sequencing with an instrument form factor.

Figure 4.1 Studio Photo

Keyboard

The number one choice for sequencing is the piano style keyboard controller. Composers are often trained to play some level of piano, and those who aren't still typically know their way around the instrument. Which is the best kind of keyboard to use? What are the major differences? These are some of the questions this section aims to answer.

Figure 4.2 Studio Logic SL88

Categories of keyboards

Assuming that cost isn't a factor, there are reasons why someone would make various keyboard selections. Perhaps they want something which has the feel of a grand piano or they frequently travel and want something that fits into their bag. Some might want a plethora of faders, knobs, and buttons on their keyboard but others want something simple and efficient.

Perhaps in order to avoid needlessly creating the ultimate comprehensive guide this should instead focus on a more realistic approach to what an imaginary reader might actually be interested in.

Figure 4.3 Realistic Home Composing Station

In summary, higher quality means something more expensive but that isn't always what serves the end goal in the best possible way. Here are some of the overriding factors to consider when selecting a keyboard for digital orchestration.

Sample select

Key-switching is regularly used by virtual instruments, where articulations and performance techniques are accessed by triggering a note in the lower or higher notes outside of the range of the instrument. If such instruments are part of the planned workstation then it makes reasonable sense to utilize a full 88-key controller. This provides access to the instrument and the key-switches simultaneously.

Figure 4.4 KeySwitch Example

Expand-ability

How many additional controller devices will need to be connected to the keyboard? Does it have ports and connection options? Some keyboards allow for a single sustain pedal, while others have enough ports to connect multiple different pedals.

Figure 4.5 Pedal Ports

Built-in sounds

Computers offer the farm to musicians in terms of available sounds and functionality. What if an idea comes and all that is needed is a keyboard capable of playing various sounds? Some composers like to turn off the computer and noodle away on a piano patch without worrying about all of the fancy sequencing tools. A MIDI controller which is only designed to send out MIDI data but has no sound engine will only work when the entire system is turned on and running. If there is a need to use the keyboard without a computer constantly attached then using a keyboard with sounds becomes more important.

Expressiveness

Figure 4.6 Author's Preferred Noodling Keyboard is a Synth

MIDI keyboards are often technological dinosaurs, with little or no innovation for decades. This is finally changing due to the development and release of MIDI 2.0, which is bringing expressiveness to keyboards across individual notes as never before. Polyphonic aftertouch (MPE) is starting to roll out across the realm of orchestration and will make waves in terms of control and performance-based sequencing. Specifically, this means that each key on the keyboard can generate its own aftertouch control data.

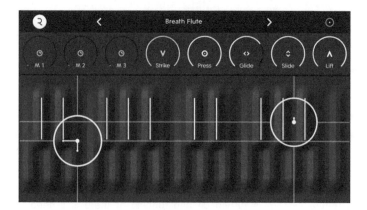

Figure 4.7 Roli App is Capable of Polyphonic Aftertouch

Keyboard alternatives

Keyboards are not the only type of controller, with various keys, wind-style instruments, MIDI guitars, and pretty much anything you can think of to make music. It isn't practical to write about each one and things are in a constant state of development.

Non-note MIDI controllers

It is possible to create a full orchestration in your software using a piano style keyboard and careful editing of each note, but the use of controllers helps both efficiency and realism in the final product. Controllers come in a wide variety formats, price ranges, and design objectives. In this section the most common types are explored, with recommendations for which are best in digital orchestration.

Figure 4.8 Modulation Wheel and Pitch Bend

Modulation Wheels

In the previous section the details of keyboard MIDI controllers showcased the near universal inclusion of pitch wheels and modulation wheels. The problem with modulation wheels often includes the size, location on the keyboard, and the inability of a wheel to have high resolution control over your data. If a wheel is all that is there then it can still be used for the purpose of modulating your performance, but other controllers can offer significantly better options. While most pitch wheels and modulation wheels can have their data rerouted, the major difference between the two typical types of wheels prevents them from being used in every situation. The pitch wheel is spring loaded so that the wheel in its resting position stays at 50 percent, and effort has to be exerted to move it in the positive or negative. Other modulation wheels are not spring-loaded move freely from 0 to 100 percent.

Figure 4.9 Joystick

When determining which keyboard to use the inclusion of flexible pitch and modulation controls is an important factor. The size and quality of the wheels can make a big difference in the ability to control the performance with the wheels. Thoroughly test out playing the keyboard to ensure the location is comfortable and easily accessible because if they aren't then it is highly unlikely that they'll be used on a regular basis. Most of the techniques described in this text require or are greatly enhanced through the use of high-quality controls. A modulation wheel certainly capable overachieving high-quality results but as is evident in the following pages additional controllers or able to enhance the workflow.

EXPERT TIP

A spring loaded wheel is most often used for pitch bend but that doesn't mean you can't use it for other data control. Converting the pitch data to control volume when sequencing an instrument's performance establishes a baseline level which can then be added to or subtracted from to make minor changes and if the wheel is released then it returns to the baseline level. This can be a timesaver when working under a tight deadline because it ensures that the instruments are all at a baseline level, which can be determined in your software.

Figure 4.10 Modifier MIDI FX

Foot Pedals

One of the most common control devices is the foot pedal. These are used for sustain, volume, patch switching, and so many other tasks.

Figure 4.11 Sustain Pedal

SUSTAIN PEDAL

The most effective type of pedal for sustain is a pressure engaged pedal which only sends data when depressed and then releases when the foot is removed. This type of pedal is commonly connected directly to a keyboard with a 1/4" TS connector and the data is merged with the MIDI output.

Figure 4.12 Expression Pedal

EXPRESSION PEDAL

This pedal type has a range of expression instead of a just an on/off like the sustain pedal. This type of pedal is often used to adjust the volume of an instrument but can control any parameter. It is also commonly used as a second pedal with the non-primary foot and can be used simultaneously with the sustain pedal. It is much like driving a standard automobile where one foot is on the gas and the other is on the clutch. This requires some practice and time to master.

Figure 4.13 Switch Pedal

LATCHING PEDAL SWITCH

This type is similar to the sustain pedal but has two positions that remain latched when engaged. If the pedal is clicked on then it remains in that state until it is clicked again. These seem to be less common now that software is capable of using switch data and latching internally, but some keyboards, guitar amps, and DAWs still find latching a useful pedal type.

Figure 4.14 Joysticks

Joysticks
A common alternative to controller wheels are joysticks, which provide much of the same functionality but with the ability to move into directions creating a host of combinations between the X and Y axes. When performing on a keyboard with the right hand, a joystick provides more control with the left because, for instance, modulation data and pitch data

could be controlled simultaneously. Theoretically this might seem like a strength but it often brings as many issues as it does advantages. It can be quite difficult to adjust modulation while maintaining a pitch adjustment. On keyboards which have added a joystick to minimize space usage, it is unlikely to be a high-quality joystick and it is therefore more difficult to achieve consistent results. To get the best results from joysticks, with optimal flexibility, multiple joysticks or a combination with wheels is required.

A keyboard such as Studio Logic SL88 has three control joysticks, with three different spring configurations. The first joystick has spring loaded functionality on both the X and Y axis directions. The second joystick is spring loaded only on the X axis. The third joystick is not spring-loaded at all. All three can be assigned freely to different parameters, although the first joystick is assigned to pitch band on the spring-loaded X axis.

EXPERT TIP

Think outside the typical volume, modulation, and pitch bend when assigning joystick parameters. One could control volume while the other controls the cut off frequency of a filter. As the instrument gets louder through joystick control, it would be easy to move the other parameter and make it brighter at the same time.

Buttons, Faders, and Knobs

Any device which is capable of sending MIDI data based on a physical movement or trigger is capable of being a MIDI controller. Even in cases when specific control data cannot be altered or changed, all input data can be modified and altered in the software sequencer.

The benefit of being able to use a wide variety of physical controllers is that many MIDI keyboards have them and are often relatively inexpensive.

Figure 4.15 Buttons

BUTTONS

Buttons tend to be binary devices which are either on or off. It is possible to use them as a control option in the same way you could use a foot sustain pedal, to switch a mode or engage a feature. The note data from drum pads can be adapted into use to control other MIDI parameters. Some pads, such as those made by Akai, are often pressure sensitive and include after touch-style pressure output for use in re-triggering notes or continuously changing velocity without taking your finger off as notes are triggered.

Figure 4.16 Faders

FADERS

MIDI faders are most commonly simple vertical or horizontal sliders which output a single stream of data. Some faders are motorized which provides useful visual and mechanical feedback when adjusting data, but both kinds are useful when controlling parameters.

Figure 4.17 Knobs

KNOBS

Knobs are circular controls and are useful for a variety of tasks. While the functionality closely overlaps with what faders are able to do, many knobs are designed to have continuous spin which is something a fader could never do.

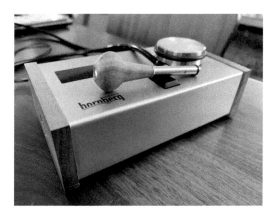

Figure 4.18 Breath Control

Breath Controllers

This type of controller uses your breath to create MIDI data. A device capable of breath control is held against the mouth and is either part of a larger MIDI controller with buttons and keys to play as if it's a full instrument or it is just a controller with the mouthpiece and MIDI connectivity. Some keyboard instruments have optional breath control adapters.

Breath control is one of the most difficult tools to master in the sequencing process. It requires skill to master how you breath and experience holding pressure over longer periods of time without being able to breath whenever is desired. In the scope of orchestration, a breath controller is sometimes an ideal partner when programming instruments which, in the non-digital world, rely on breathing such as woodwind and brass instruments. Imagine being able to sequence a flute part with phrasing that is both realistic and natural because it was created using a breathing process. This also frees up a hand in the recording process, which with a wheel or joystick would otherwise be engaged. It could also be used in addition to a separate controller, providing the opportunity for a more complex control scheme.

While the data coming from a breath controller can be assigned to any parameter, it is limited in other significant ways.

- It isn't possible to use multiple breath controllers simultaneously, unless you have more than one set if lungs.
- If you run out of air or breath too much it is possible to get dizzy and potentially pass out.
- It can take a lifetime to master wind instruments and the breathing required to perform on a professional level, which is the same for this type of controller.
- Pressure is required for data creation and breathing in for more air cuts the pressure completely.

The technique needed to effectively harness the power of the breath controller is split into two primary categories. The first is to push air pressure against the controller while essentially holding air in the mouth and lungs. The second is to breath out past the controller while ensuring pressure still hits the control interface.

BREATH HOLDING

The first is essentially holding one's breath which involves releasing all of the air and sucking in new air during breaks in the music to allow the body to continue to get the oxygen it needs. This is different than most non-digital wind instruments because very few of them involve holding air inside when being played. Exceptions include instruments like the oboe, which have a narrow passage way allow for less air to pass than instruments like the trumpet or trombone. This method is often the way that most would use when first trying a breath controller because it feels natural to push air directly into the interface which doesn't let air pass but you'll hear the effect of the data on the sound of the software instrument. This method almost always produces the most dynamic results but is hardest to manage due to hoarding of oxygen as it turns into carbon dioxide in the lungs.

Figure 4.19 Neck Loop

AIRFLOW

The second method is to let some air out as you blow into the controller. As long as enough pressure is created in the controller, then this is an acceptable method and has several benefits. The first benefit is that the lungs will be nearly or completely emptied while performing and thus allow for a quicker refill in the breaks. The second is that it feels closer to performing a wind instrument, where airflow helps the musician gauge pressure. The one variation on this which rarely produces acceptable results is breathing out through the nose while also blowing the controller through the mouth. This prevents enough pressure from being created in the breathing process.

A well-designed breath controller system offers more than simple pressure ana-lysis and MIDI data output. It is possible to program hardware to account for breaths, added decay time when pressure stops, and other modified data options to compensate for the inherent shortcomings of measuring breathing pressure. The Hornberg Research HB1 MIDI Breath Station is one of the few comprehensive options with extremely high-quality com-ponents and the ability to interpret incoming breath pressure and convert it into highly ex-pressive control data. This particular device is showcased further here and in Chapter 6.

Figure 4.20 Leap Motion

HANDS-FREE CONTROL

This type of controller has a varied history with devices coming and going and seemingly no consistent front runner in the marketplace. For a number of years, the Leap Motion device had the most promise and it used infrared and intelligent video analysis to create a decent mapping of a performer's hands. The device is currently still available and there are still software MIDI converters for it, but development seems to have stalled. The X-Box had a hands-free controller which has since gone away. For the purpose of this book we use the Leap Motion to demonstrate the possibilities, while exploring all of what the future may bring in Chapter 7.

Figure 4.21 GarageBand Face Control

App Controllers

Both iOS and Android have MIDI controllers which are two dimensional or rely on movement sensors but are widely creative and functional. iOS GarageBand has one of the only facial recognition MIDI controllers currently available and hopefully it is expanded to control additional parameters. Connecting an iOS device into a production/composition setup is easier than ever before with the integration of the Inter-Device Audio + MIDI (IDAM) protocol released by Apple. There are also a large number of WIFI, Bluetooth, and wired options for communicating between devices. Some of these are explored in the examples later in this text.

Figure 4.22 Logic Pro X App

It is hard to predict which technology will endure in this modern technological world, and iPad apps have proven a difficult market to navigate. For a few years after the original iPad was released, there was a large number of developers creating music control apps and it seemed like this would continue to flourish. The number of powerful apps has remained wide, but the landscape has shifted. Developers have let a lot of apps fall into disrepair, with no updates for years and without any replacement from other companies. The apps developed by companies with an existing customer base and other significant products seem to continuing app development but without the expectation of revenue earning. Logic Pro and Avid's Pro Tools both offer companion apps but without cost. Some of the classic apps such as Lemur, which is an amazing controller app, still functions but hasn't had an update in several years.

Figure 4.23 Lemur App

Software set-up

In this section the connections and settings are explored in order to provide the most current options available. With the wide variety of options, the best bet is to research the individual piece of gear to ensure it is compatible and then contact the individual company for trouble-shooting if required. The net being cast here, so to speak, is to share the most common pitfalls that someone new to digital orchestration might not even be aware of when first creating a production setup.

Latency

There is delay in all digital systems, which is actually not too terribly different from the delay when an orchestra performs. The conductor waves her baton and it takes a moment for the musicians to follow it and play. Wind has to leave the lungs and travel out of the mouth and into the instrument. Things take time.

There biggest difference is that an orchestra is made of musicians who can respond to delays intelligently and adapt to remain "in time" with the group as a whole. Their ears provide a link from the acoustic waves to an intuitive, perhaps even sub-conscious, analysis of what is happening and a smooth trigger to speed up, slow down, or maintain the current speed. See Chapter 5 for more detailed information about latency and the buffer settings involved with recording MIDI.

There are solutions which can be adopted in the case of a large mix where you also need to continue recording:

1. Use a low latency mode if available in your sequencer or in the instrument if external. This deactivates portions of the mix until a set amount of latency can be maintained.
2. Create a new project with an export of the large mix and then add the new part. This removes all of the portions of the mix which are causing the delay and creates a low latency option. It is a pain to have to switch projects but still a viable solution.
3. Draw in the new part instead of recording it live. This eliminates the issues cause by delay.

No matter what, learn where the buffer setting is in the software and check it on a regular basis, especially before recording.

Connections

There are a number of different ways to connect external controllers to computers and it's worth exploring so that the different options are clear. The goal is to explore these connections as an overview but not to go in too deep because understanding how the individual protocols work only helps in a few key instances, which will be covered.

Figure 4.24 USB Ports

USB

There are currently a number of USB types which are available for use with MIDI. A music production studio could easily be using all forms of USB, requiring adapters, USB hubs, and swarms of cables.

Benefits
- Near universal.
- Most devices work without installing software.
- Easily expandable.
- Includes power for external devices.
- Reliable.
- Backwards compatible.

Weaknesses
- When a connection doesn't work it is difficult to troubleshoot.
- Limited bus power.
- Apple keeps changing connectors, which get expensive.

Troubleshooting
- Unplug and re-plug.
- Power cycle.
- Look for updated software/drivers.
- Explore online forums.

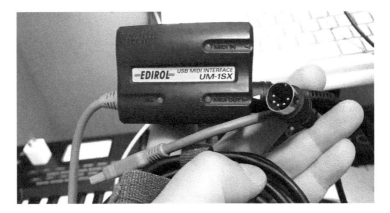

Figure 4.25 MIDI Cable

5-Pin MIDI

When MIDI was first released it utilized a 5-pin DIN connector well suited for low voltage signal transference. As an option, this original connector is often still included with hardware controllers, although considering only three of the pins are ever regularly used, it is one of the great resource wastes of all music technology.

Benefits
- Continuous use of the protocol means the oldest MIDI technology is still available for modern music production.
- Rarely fails, even after 35 years.
- Simple connections requiring no software, or even computers. Multiple devices can be daisy chained together.

Figure 4.26 Daisy Chain

Weaknesses
- The format sends data one piece at a time which translates into small, yet additive timing errors.

- The technology also uses photo resistors at either end of the connection which adds latency.
- MIDI connections use a cable for each direction of communication, which also send only 16 channels of musical control.

Troubleshooting

- Use a MIDI monitor app to track data.
- Try an alternate cable to see if it has failed.
- Read the manual to figure out where to turn on/off local MIDI.
- Some controllers have 5-pin inputs/outputs but require more than USB bus power to operate. Ensure that there is a wall plug as well.

Figure 4.27 Bluetooth MIDI

Bluetooth MIDI

This wireless connection type uses the existing Bluetooth protocol to send MIDI back and forth. While it is a short-range option, it seems to perform well and reliably.

Benefits

- Removes the need for additional cables.
- Frees the performer to move away from the computer.

Weaknesses

- Initial setup can be tricky and it isn't recommended to switch devices between computers regularly.

Troubleshooting

- Unpair and re-pair.
- Power cycle everything.

WIFI

Wireless WIFI MIDI, one of the least used formats, won't even get a description here as it doesn't apply to controllers and is mostly used when computers are communicating together.

Figure 4.28 IDAM+ Example

IDAM+

One of the exciting newer formats is the Inter-Device Audio and MIDI protocol which is a part of the OS X operating system. It allows an iPhone or iPad to be connected to a laptop/desktop computer and it sends MIDI back and forth, along with audio from the mobile device, into the computer. The majority of mobile app instruments are more suited to pop/rock/EDM styles, but there are a few orchestral instruments being developed which can add to the overall musical toolkit.

Benefits
- Super easy to setup in the Audio MIDI Setup utility.
- Allows mobile devices to be incorporated into the production process.
- No additional cables (aside from the included charging cable) are required.
- Audio remains in the digital realm the entire time.

Weaknesses
- Mac only.
- Apps are hit/miss when it comes to quality and there are no standard expectations.
- Requires a cable.
- Only one instrument on the mobile device can play at a time unless other apps are utilized to allow multiple apps to play simultaneously. This still only allows a stereo feed to be sent from the device so all sounds will be mixed together before coming into the workstation.

Troubleshooting

• Very few issues but occasional a restart required and sometimes it helps to close all open apps on the mobile device.

Figure 4.29 Touché Controller

Most relevant examples

In this section four different scenarios are explored to showcase what is possible with various control situations, with a special focus on tools that are especially valuable in the orchestration process.

Scenario 1: Touché

The Touché from Expressive E is one of the most incredible controllers for use in orchestration. It is big, with enough mass to feel weighty, but capable of genuinely subtle and sensitive control. The device uses and ingenious system of multiple rubber shapes inside to give movement in four directions, with additional movements up and down for the separate halves. The Touché has three different types of connectors; it has USB, 5-pin DIN, and control voltage (CV). In this example, the Touché is connected via USB to a computer running Logic Pro X. The controller is used in addition to a Studio Logic 88-key piano keyboard. Sequencers are able to work with multiple data sources simultaneously, with the keyboard sending note data and the Touché sending expression data.

Figure 4.30 Touché Guts

There are several common ways to incorporate such devices into a project. The first involves performing both at the same time and having the software merge the streams together. The second option is to play the notes first and then record a second pass with the Touché adding volume data or other control. There is something elegant to performing both together, but it takes practice and control. Without the proper skill, recording the parts separately is recommended.

Figure 4.31 Playing Both Together

The Touché has a fixed CC configuration, using CC 16, 17, 18, and 19. These don't align with typical control destinations and so the software needs to be set to convert the data. In Logic Pro there are MIDI FX which do a variety of things. The Modifier is perfect for this situation because it can easily be set to change CC 16 to any other MIDI parameter or plug-in control. In orchestration it is not common to need all four axis directions to manipulate the performances, which is why the developers used 16–19 due to their infrequent application.

The first step is to ensure the Touché is attached, powered up, and sending data. Because the Touché has onboard memory it can track what you've done in the past, but often in cases like this with Logic then it might make the most sense to reset the device you holding both arrow buttons when powering on.

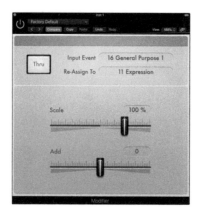

Figure 4.32 Modifier FX

Next, engage the Modifier FX and set it to change CC 16 into CC 11. Turn off the THRU function so that no CC 16 continues through to the instrument. Add an instrument to the track and then test out the Touché to make sure it is controlling the Expression parameter. Every patch is going to react differently and have different features, so it is important to be prepared to adjust the Modifier as needed. Once the patch is selected then adjust the sensitivity knob on the Touché until the sound is natural when being controlled. Record the instrument's part first and then do a second pass with the Touché. It's critical to set the MIDI function in Logic to merge when recording over existing data, otherwise the notes would be erased when recording the second pass.

Figure 4.33 Merge Function

After both portions are recorded then it is time to listen carefully to the performance and to begin the editing process. Perhaps it sounds perfect and needs no additional work, but even the best performances often need a little housekeeping. Open the piano roll and then the MIDI automation lane. The view may have to be switched to see CC 11, but then the data can be edited to fine tune how the instrument sounds. There are a few different tools for MIDI automation editing, but the easiest is using the pencil tool to draw in changes to the recorded curve. If the curve isn't what you want and perhaps it makes the most sense to re-record the Touché, then the merge mode would have to be deactivated to avoid a huge data mess. It would record new data points but leave many of the original ones in place, creating a zigzag result.

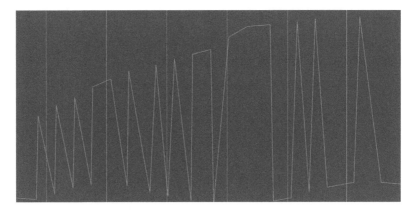

Figure 4.34 Zigzag Result

In some advanced patches there are features which are well suited to control with the Touché. Many of the instruments have versions which are tied into the Modulation control to fade between different dynamics or articulations. One great example is a woodwind sound which transitions between note lengths from sustain to staccato and is controllable by default through the Mod wheel. Using the Modifier FX, the Touché data can be converted to the same as used by the Mod wheel, except instead of the same, as in the example above, the data would come from the back half of the Touché. The front half would still be used to control the volume or expression of the instrument and then the back half would control the length of the notes and articulation. The side to side motion of the Touché could then also be assigned to pitch data, giving extensive and comprehensive control over the instrument.

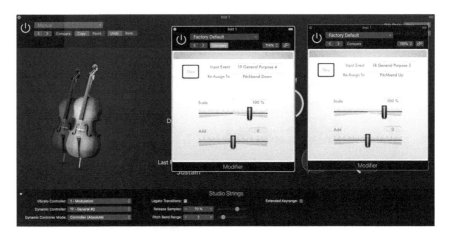

Figure 4.35 Advanced Modifiers

Recording four different types of data simultaneously is very powerful and efficient but there are some side effects which make it difficult to be the golden bullet.

The Touché is high quality and built well, but it is still easy to move it in such a way that it triggers in all directions. This means it might be the intent to push down the front half, but it slightly moves side to side as well. The data created is also four times as much as when recording one parameter. Data management becomes even more important in the editing phase and keeping track of which stream is being edited could be the difference in mistakenly editing the wrong parameters.

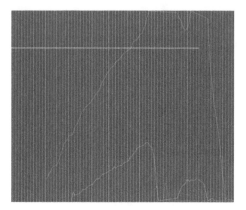

Figure 4.36 Multiple MIDI Automation Lanes

This isn't the only device capable of working like this, but is certainly one of the most powerful ones. It is sturdy and precise and feels more like a musical instrument than a glorified computer mouse. The key, as always, is to use it often enough to get used to it and master its feel and functionality.

Figure 4.37 Hornberg hb1

Scenario 2: Hornberg HB1

The second scenario showcases the HB1 from Hornberg Research, which is one of the leading breath control units. In addition to having a top-notch physical design, it also has a very intelligent software design. While there are some similarities to the Touché scenario above, all of the pertinent information for set-up in the hardware and software are explained in full for the HB1 so that it stands alone as a resource for those interested in using it or other breath controllers.

Figure 4.38 hb1 Connections

The unit has two connection types, with USB being the primary option but it also includes three MIDI connections, which includes one input and two outputs. The unit is powered by USB, although it can work with or without a computer. In stand-alone mode it must still be powered by USB but can take incoming MIDI data and add in control data resulting from the breath input. This makes the device perfect for live performances.

Figure 4.39 Mouthpiece

There is a mouthpiece which is what the performer blows on, a tube which leads to the breath module, and then a cable which leads to the main unit. It is possible to hold the mouthpiece in the mouth, but it is recommended to use an included neck brace to hold the mouthpiece in front of the face and then blow when desired without having to focus on also holding the mouthpiece up.

The main unit is a button and a knob, which are used to set the control data parameters. Here is a list of available parameters:

Pr: Presets 1-25

dr: Drive (sets pressure sensitivity)

oF: Output offset

At: Attack

bA: Boost attack

rL: Release

br: Boost release

Ll: Output limit

rd: Reduce data

CC: Continuous controller

Ch: MIDI Channel

CP: Combine preset

Figure 4.40 Data Screen

The button is used to select the parameter and then the knob is used to change the value. In the Touché example above, the output is a set of fixed CCs which are then changed in the software, but on the HB1 the CC is assigned before sending data to its destination. After attaching the HB1 to the workstation, the CC parameter can then be assigned to the Modulation CC 1, the Volume control CC 7, or the Expression CC 11. Unlike the Touché example, it is more intuitive to immediately use the breath controller simultaneously while recording MIDI from a keyboard, but there are other significant issues described earlier in this chapter when discussing breath controllers – they can be difficult to master in terms of performing due to the need for developing breath control.

Figure 4.41 East/West Instrument

In this scenario the objective is to perform one of the opera patches from East West's Composer Cloud. The CC parameter is set to 11, which works well to control the loudness of the instrument except when the performer needs to take a breath. Certainly, there need to be breath spots in order to make the performance seem natural, but it's a different type of breath when using the HB1 and it doesn't match an opera singer's breathing. To give the MIDI performer maximum flexibility, the HB1 rL parameter is engaged, which adds slight sustain and then a period of release after the breath pressure is released. This means additional quick breaths can be snuck in during recording to keep the vocal instrument going until a longer space is appropriate for a full breath. In order to make it sound like the virtual voice is breathing, it makes more sense to edit them manually and since it is harder to maintain constant breath throughout, it is a huge time saver to let the rL maintain a constant signal.

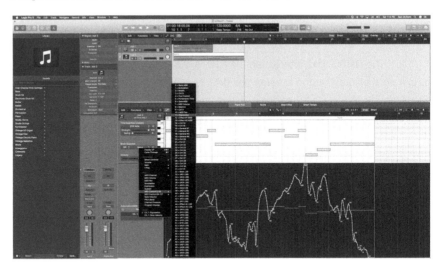

Figure 4.42 MIDI Automation View

To edit the data view the MIDI in the piano roll and open the MIDI automation view. Navigate to where you want a breath and switch the view to show CC 11. See what the data looks like at the spot and determine if it can be reduced to provide space for a breath.

The MIDI connections are useful in a very different way, with the option to work outside of workstation software. There would need to be a MIDI source and destination for this to work, although it is technically possible to have both in the same keyboard workstation. The originating MIDI keyboard is the source and it must connect via MIDI cable to the HB1'S input port. The breath data is captured and then merged with the input data, creating a new stream which is sent on to a destination instrument. The destination could be a sound module, a separate keyboard, or even the same keyboard as the source. If the plan is to use the originating keyboard then Local MIDI should be deactivated so that pressing a key doesn't trigger the sounds on the instrument but still sends MIDI data out. The MIDI is then sent to the HB1 where the data is merged with the breath data, which is sent out and back to the original source. That new MIDI is able to trigger sounds on the instrument even though Local MIDI is turned off. The end result is a performance which is the result of two sets of data but one instrument.

Figure 4.43 Negative dr Data Value

The last thing of interest on the HB1 is the ability to invert the breath module output so that breathing in creates the data normally expected when breathing out. Some people prefer this breathing action in some circumstances and it is easily achievable by setting a negative Drive value on the HB1.

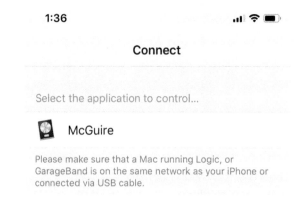

Figure 4.44 Logic Remote App

Scenario 3: Logic Remote app

The third scenario looks at using an app to give additional control over the orchestration process. In this situation the app is Logic Remote which is paired perfectly with Logic Pro because it is from the same company. When looking for good control apps, look at options from the same developer as the workstation because they often provide paired solutions. The Logic Remote app is well suited for orchestration because in addition to normal mixing features, access to program shortcuts, and the ability to manipulate effects, the app also has the ability to act as a MIDI controller. It can do more than just trigger notes, but also has velocity sensitivity and can control instruments in a variety of ways.

Figure 4.45 Connection Screen

Connecting the iPad app is simple and only requires that both the iPad and the host computer are on the same network. If Logic is running then it shows up in the app as a possible connection, which is easy to pair with the click of a button. There are several interesting ways to use the app in the orchestration process.

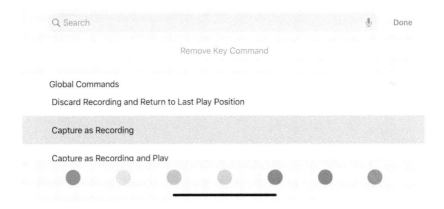

Figure 4.46 Assigning Menu Options

First, it's possible to assign any menu items and functions to the shortcut page in the app. This makes using some of the various tools even easier and bundle options together into a single view. Second, it's easy to adjust instrument settings even from third party sound libraries. The more that can be done without the mouse the better in terms of efficiency, and since a mouse can only accomplish one thing at a time but the iPad has a multitouch screen which makes it possible to manipulate more than one parameter at a time. Finally, there are a variety of instrument control screens which are designed to mimic real instruments. There are drums, guitar strings, and keyboard instruments. While GarageBand for iOS has a full set of string instruments with string-like control that is actually very similar in appearance to real string instruments, but in practice it is very dissimilar to actually playing the original instrument. The Logic Remote app has many of the same features but still not the string control option.

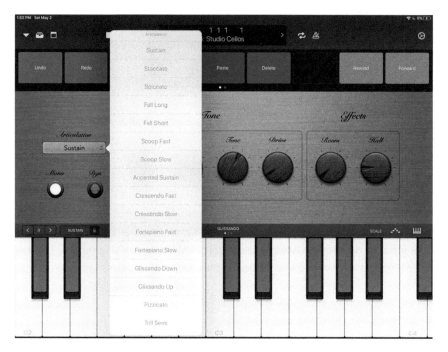

Figure 4.47 Studio Strings

The core part of this scenario explores the benefits of the iPad format, specifically the things which are available in the app which are not easily found anywhere else. The first step is to add an orchestral instrument using the app, which is easy to do and, in this case, is the Studio Strings built-in instrument. The reason for this is that it has additional features already programmed, including extensive articulation access. It is possible to play notes with one hand and switch between playing styles with the other hand. The changes are recorded along with the note data.

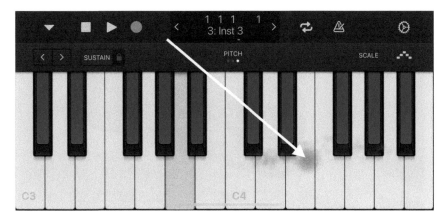

Figure 4.48 Pitch Control

Another feature available in the app is the ability to change how the keyboard reacts when utilized. There are three modes which are glissando, scroll, and pitch. Glissando plays the notes as the finger is swiped across them. In scroll mode the keys shift through the various ranges as the finger is swiped across them. The last mode, pitch, is unique to working on the screen where a note is pressed and as the finger moves along the keyboard the pitch is shifted up or down. This means one finger can play notes and adjust pitch bend, while the other is selecting the articulations. The pitch mode is useful for creating custom vibrato, which isn't as easily done in any other way. All of this doesn't even account for pedals, which add additional control or a breath controller. One person can do a lot of music creation with these tools.

Figure 4.49 iPhone Drummer

The last part of this scenario is the use of the iPad or even iPhone in the creative flow. Sitting in front of a computer screen is something that is part of the routine, with the computer keyboard and mouse. Sometimes it is better to sit in front of a keyboard and nothing else. In this case it is okay to have a musical keyboard aimed in a different direction or even in a different part of the room. Load up the control app and run the session from it instead of the big station. Logic Remote on a phone is barely noticeable and doesn't distract from the musical experience.

Figure 4.50 Leap Motion

Scenario 4: Leap Motion

In this final scenario the Leap Motion device is explored in greater depth, showcasing how a touch-free device is able to help in the orchestration process. The original Leap Motion company was in business from 2010–2019 and was sold to UltraLeap who has continued to make the product available. The Leap Motion is an optical hand tracker that works amazingly well at translating hand motion into a data stream. It is unclear if new apps will be developed for the Leap Motion but currently there are a handful of options to translate hand movement in the air into useable MIDI data.

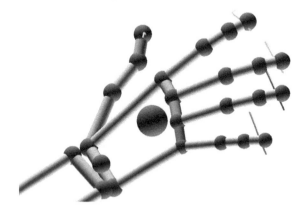

Figure 4.51 Leap Motion Visualizer

The primary utility used in the scenario is called Geco MIDI, which creates 40 data streams from the movement possibilities of both hands. Each stream can be assigned to any of the MIDI Continuous Controller assignments, which make this a powerful way to control volume, modulation, expression, or anything else.

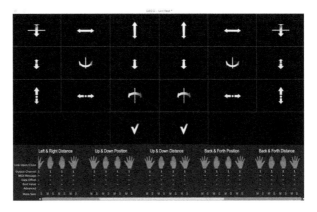

Figure 4.52 Geco MIDI App

In order to make this work the Lapp Motion software and the Geco MIDI app both need to be installed. Inside Geco there are a number of boxes which are attached various hand motions, both with fingers extended and held together. For this example, the left hand with fingers extended is selected, with the up and down motion being assigned to CC 7 for volume and the left-hand rotation movement being assigned to CC 16 as general assignment. The reason for assigning the rotation movement to CC 16 is so that it can be used to control an equalizer effect in the workstation to increase the brightness of the instrument separately from the volume.

Figure 4.53 Assigning the CC Value

A trombone player can play dark or bright somewhat independently from its loudness or softness. In Logic Pro the CC 7 parameter is automatically passed on to the instrument's volume but CC 16 has to be mapped to the equalizer. After adding the equalizer to the instrument track, the mapping of CC 16 is accomplished in the Smart Controls pane, where the gain of a high shelf band is attached to the CC. The end result is that when the left hand is held over the Leap Motion and moved up and down then the volume is adjusted accordingly. This is a very intuitive way to add volume control. The rotation motion is able to add brightness to the sound via the equalizer – when the hand rotates it gets brighter. A performance could include a trombone sound getting louder but maintaining the same tone, or the tone could brighten at a similar rate to getting louder. Technically it could also get brighter without getting louder.

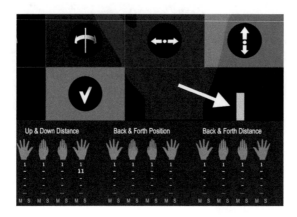

Figure 4.54 Geco MIDI in Action

One more thing to consider when setting up the Leap Motion with Geco MIDI is that a baseline CC can be set and also the behavior of what happens when the hand moves out of range. The benefit to setting a base level is that the instrument can be played before engaging the hand above the Leap Motion. The behavior dealing with the hand leaving the acceptable range is also important because it could return the CC to the base range or it could be set to remain at the last known number. There really isn't a wrong answer, but it is more important to be aware of what the setting is.

Figure 4.55 Complex Setup

Another example of the Leap Motion in action uses multiple data conversions to change hand motion into note data, to set a tempo via the tap tempo function. Is this the easiest way to do this? Not by a long shot, with it being much easier to simply tap a key on the keyboard. However, there is a certain aesthetic to "conducting" the tempo in the air as it might be done in front of an orchestra.

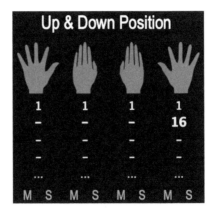

Figure 4.56 Setting Geco to CC 16

The first step is to set up Geco MIDI so that the vertical hand motion of the right hand is attached to CC 16. Then attach the CC 16 to the Tap Tempo key command, in the Key Commands Edit menu.

Figure 4.57 Tap Tempo Key Command

There is a Learn option on the right side of the dialog. Simply enable the Learn option and then move the hand above the Leap Motion.

Figure 4.58 Sync Settings

The tempo settings in Logic also need to be set to allow for manual tempo control, sync, and to allow Logic to record an input tempo.

Figure 4.59 Tempo Interpreter

Last, engage the Tempo Interpreter from the contextual menu on the main window. Now it is ready for tempo recording. Engage the record button and let the four counts happen before moving the hand up and then back down. The tempo will track a data point on the way up and the way down, so keep the motion steady and at the desired tempo. The movement would have to be very consistent in order to create useable tempo information but is a viable option if it feels more natural to wave the hand. Is the setup required for this insanely complicated? Undeniably yes.

Figure 4.60 Recording the Wave

The Leap Motion is very flexible and can be attached to so many different parameters. It can move through the same parameters which the mod wheel controls, it can adjust the length of the release parameter, it could be attached to the pan knob on individual tracks, or a thousand other things. The breath controller is well suited to instruments which rely on breathing, the Touché is amazing for a variety of tasks with its multiple axis directions, but the Leap Motion can track 40 different things simultaneously and without ever requiring a physical touch. Virtual reality and augmented reality are the only technologies currently available on the horizon which could possibly add more control in the future and it is only a matter of time before someone makes a commercially available solution to control music in the virtual realm.

Summary

The goal of this chapter has been to explain the options available for creating and manipulating MIDI data through the use of controllers. Realism in a digital orchestration comes not from the source instruments alone, but in how they are given life through the addition of expression control and other sculpting tools. There are more than enough tools described here to use and probably more than any one person is capable of ever using. In the end, we recommend picking the tools which align with personal working styles and to develop skills using them. Like any instrument, becoming proficient takes time and consistent effort. Even the most intuitive controllers still require practice.

Chapter 5

MIDI techniques

MIDI is half of the mountain to climb when mastering digital orchestration. Getting to know the available sounds and what they are capable of is certainly important, but this chapter focuses on what sequencers are able to do when controlling the sounds to make them sound realistic and/or fulfill the goals of the orchestration.

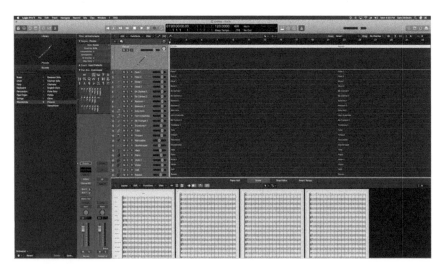

Figure 5.1 A Fresh Orchestration in Logic Pro X

This isn't going to be another book on what MIDI is and how to use it. For all practical purposes it is assumed that each reader is already working at a high level of sequencing but perhaps needs additional information to start down the path of orchestration, to fine tune their sequencing skills, or to simply learn new MIDI techniques applicable to any genre. If rudimentary skills are lacking then perhaps study the basics of MIDI in one of the other books available from Routledge, such as *Modern MIDI*.

Recording MIDI

There are two important phases of working with MIDI, starting with the input phase where MIDI is recorded, drawn, or generated into a sequencer. The second phase is editing and manipulating the data.

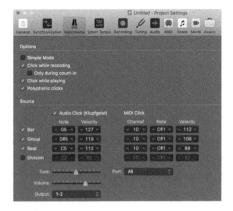

Figure 5.2 Advanced Metronome Dialog

Click-track

It used to be that one of the most important steps before recording MIDI was to set up a click track and tempo map. Not only would this make editing MIDI easier, it would also enable features such as quantization and MIDI effects sync. Updated technology has made it possible to accurately record MIDI without previously mapping tempo, and this technique is covered later in this chapter.

Figure 5.3 Tempo Map

A tempo map is used to match music with picture when composing for video or can be used to add simple realism to any orchestral project, which often have tempos that fluctuate. Creating a map is accomplished in a sequencer's global tempo track and can often be created using a drawing tool, a tap to tempo function, or some other tempo feature.

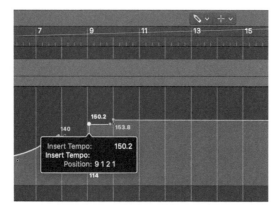

Figure 5.4 Drawing the Tempo

Drawing tempo

Drawing a tempo map is easy enough but there are some specific things to consider. Firstly, unless there is a clear reason to have a fluctuating tempo in place before recording, it is acceptable to do it before or after the recording of MIDI. One reason to have it in place first would be to match a cue with specific video timings, but, even then, it is possible to play to a strict tempo and move the timing later.

There are several types of tools used when drawing tempo information, those which allow free-form drawing and those which draw preconfigured shapes such as lines, curves, or other options. A pencil tool is relatively hard to use, given the difficulty of creating consistent changes over time. A line tool or a curve tool is more likely to be useful when creating a custom tempo. Using a sequencer with comprehensive tempo tools is critical when composing for video and simply drawing in changes is rarely accurate or efficient enough for regular use.

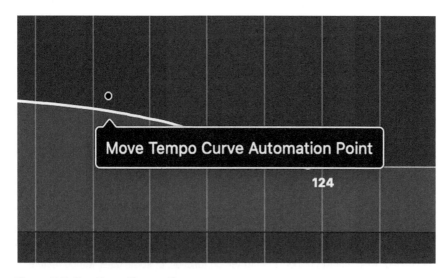

Figure 5.5 Creating a Tempo Curve

Tap tempo

This is a favorite feature for composers when setting a tempo for video work. It is most common to set the tempo using tap before any MIDI has been recorded but it is possible to do it before or after. Tap works by setting a MIDI trigger, such as a note, and then tapping that trigger as the video plays. There is typically a period of rehearsals to get the timing right to match the scene in terms of overall tempo speed and then to make sure the selected tempo works at transitions. By tapping the tempo, it becomes a performance which has the ability to make the tempo flow more naturally, with less effort than manually creating all of the changes. If the tempo is tapped, and is close, then it is possible to manually edit and fix the remaining inconsistencies.

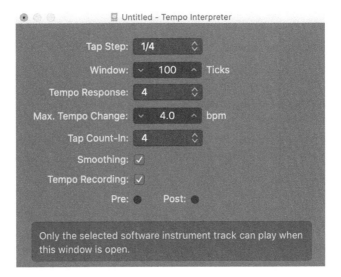

Figure 5.6 Tapping the Tempo

Tempo interpretation

An alternative to tapping the tempo is to use an interpretation tool which can analyze audio or MIDI and build a tempo around what is already there. One example of this is when a musician puts the sequencer into "listen" mode and plays a MIDI keyboard piano part without a click track. During the performance and finalized afterwards, a tempo map is created based on the performance. The analyzer uses musical patterns to create the tempo and bar structure. While it isn't guaranteed to be accurate, experience has shown a high percentage rate of success is possible.

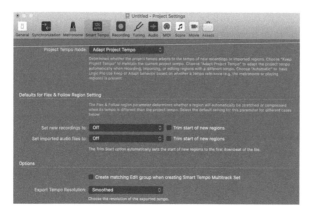

Figure 5.7 Adapt Tempo

Instead of performing live it's also possible to record audio or MIDI and analyze it later. Using a basic microphone, a composer could record themselves tapping their finger on the mic or on a different surface. The resulting audio would then be analyzed for tempo information and used to create a tempo map.

Tips for successful tempo interpretation

- Play a part which has a constant rhythmic element. If there is a held-out chord then continue to play an 8th-note or 1/4-note rhythm so the analyzer has data to analyze.
- Don't make sudden tempo changes. It is more successful to stop at the change and then restart with the new tempo.
- While it doesn't make sense to use a click track when using a real-time interpreter, it helps to set up a one or two bar tempo-based loop which is put in a track and is listened to before starting to perform the MIDI part. This establishes the tempo in your mind and helps set a foundation for analysis success.
- Using a tempo interpretation tool will not fix a poorly played musical performance. The recording needs to be fairly accurate for the process to work.

Figure 5.8 Tapping a Microphone to Record a Tempo

Demonstration

Figure 5.9

> Step 1. Create an audio track

Figure 5.10

> Step 2. Record arm the track

Figure 5.11

> Step 3. Enable Adapt Tempo

Figure 5.12

> Step 4. Engage the Transport

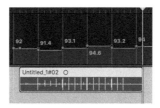

Figure 5.13

> Step 5. Tap on the microphone

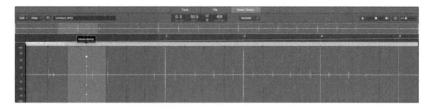

Figure 5.14

Step 6. Verify the proper starting point

Figure 5.15

Step 7. Start recording other parts with the metronome engaged

Other tempo tools

Various sequencers have tempo tools which all attempt to accomplish the same things as drawing, tapping, or interpreting the tempo of a performance. Some methods use markers created from video frames, dialogs that create smooth tempo changes, and tempo presets which pull from existing stockpiles.

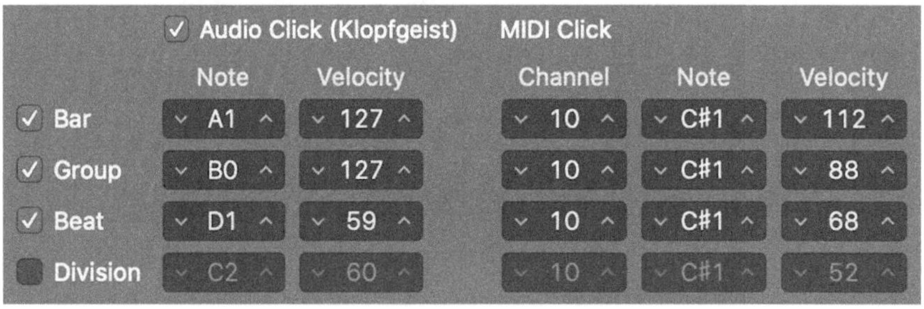

		Audio Click (Klopfgeist)		MIDI Click		
		Note	Velocity	Channel	Note	Velocity
✓ Bar		∨ A1 ∧	∨ 127 ∧	∨ 10 ∧	∨ C#1 ∧	∨ 112 ∧
✓ Group		∨ B0 ∧	∨ 127 ∧	∨ 10 ∧	∨ C#1 ∧	∨ 88 ∧
✓ Beat		∨ D1 ∧	∨ 59 ∧	∨ 10 ∧	∨ C#1 ∧	∨ 68 ∧
Division		∨ C2 ∧	∨ 60 ∧	∨ 10 ∧	∨ C#1 ∧	∨ 52 ∧

Figure 5.16 Setting the Subdivision

Recording to a click

Once a tempo map is established then it is time to record. With tempo changes it can be a hurdle if the subdivision isn't set to a small enough pattern. If the click is set to play only quarter notes then it may be difficult to play accurately when the pace slows or quickens. When recording to a tempo with significant changes, use a subdivision which is shorter than the shortest division in the music. If the part has only half notes and quarter notes, then set the click to eighth notes. If there are eighth notes, then use sixteenth notes. Any division shorter than sixteenth notes is excessive and not useful.

Figure 5.17 Adjusting the Click Sound

Click sounds

The default click sounds in most workstations are sometimes thought to be pretty mixed in quality but in fact are often functional in terms of being heard through the mixture of sounds. Their sounds are pleasant because they need to cut through the frequency range. Other sounds can be selected based on personal preferences, including realistic drum sounds, loud sounds, or even mellow string sounds. After selecting the sounds, then decide on a count-in length.

Figure 5.18 Headphones are Optional When Recording MIDI

Monitoring requirements

One benefit to working exclusively with MIDI is that there is no bleed from microphones and so headphones or speakers can be used without worrying about audio feedback. The click can be turned up as loud as needed and it won't be recorded into the MIDI tracks. The most important need is to be able to hear the nuances of the instruments while recording. A full orchestra is recorded simultaneously, with each musician worrying about their individual part while listening to the group as a whole and following the conductor. Recording an instrument at a time using MIDI is an entirely different prospect due to the nature of having to start with one instrument and a click track. Instead of having the entire ensemble to play with, the first instrument is played all alone. It is possible to successfully record each part one after another, but often a great amount of tweaking, re-recording, and patience is required to accurately reproduce an ensemble which has the ultimate advantage of playing together.

Using high quality headphones or speakers is very helpful as the instrument count increases so the details can be heard. The subtleties of orchestral instruments make adding layer after layer very difficult, unless it is possible to clearly hear how the parts fit together. For this reason, it is important to record the instruments in an organized way, with a specific plan in place, and using quality monitors. Without an acoustically designed room and good speakers, headphones are the best choice when recording. Headphones provide a microscope-esque environment for hearing the various parts, while simultaneously blocking out surrounding sounds.

Figure 5.19 A Headphone Tracker Can Create a Virtual Studio

Each piece of music is different and will need a different plan for the order of recording, but there are some tips to consider which apply more generally across multiple styles and instrumentations. The first phase is to set the foundation for the section. Some start with a piano reduction of the orchestration, but it is possible to use other individual instruments as the base layer. The key is to establish a structure in the same way scaffolding is used in construction. It would be difficult to start with an interior viola part or a sporadic oboe part which comes and goes. The strength of starting with a piano reduction is that there would be a continuous structure to work with as parts are added. If individual parts are used as the base layer then consider using instruments which represent the root of the harmony and which play consistently throughout the entire section.

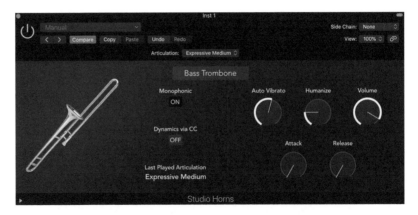

Figure 5.20 Low Brass are an Example of a Foundational Section

After a first pass is recorded, then instruments are added in an order which is logical. Focus on instruments which double each other, filling in holes as the orchestration builds. Musically, there are very few wrong or right answers in how to decide the order of adding things but as tracks are added, the processing resources of the computer also add up. The areas of highest concern are added latency and processing needs. When the processor reaches its limits there are clear alerts or signs but latency issues are more subtle and damaging to the recording process.

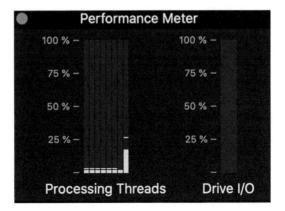

Figure 5.21 CPU Usage Meter

Latency

The more tracks, instruments, and effects which are utilized then the more work the computer has to do which results in potential latency. Latency is another word for the delay created in the audio system. Perhaps some of the effects being used add delay in order to accomplish their functions or the buffer size has to be increased to compensate for higher power needs. While it is possible to have a very powerful computer with extreme amounts of RAM and 20+ cores in the processor, which results in high track counts and room for a lot of effects, most systems in current use will have latency added as the project gets bigger.

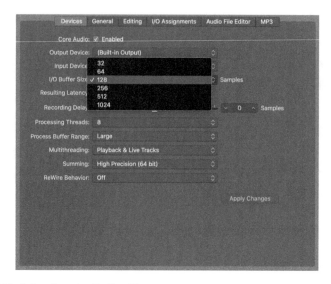

Figure 5.22 Adjusting the Buffer Size

Buffer size

Every software workstation has a buffer setting, which adds a delay of a specific number of samples in order to give the software the ability to do more. In spite of being able to travel to Mars and create 3-D digital worlds, audio software designers haven't been able to program workstations which can automatically adjust the delay involved with various tasks. This means that the buffer size has to be manually adjusted in the various phases of production, with short times during recording and longer times as the mix gets bigger and bigger. The biggest problem is that once the buffer size needs to be raised after resources run thin, that it is difficult to reduce the buffer size later. If a project has 50 tracks of instruments with multiple effects and a maxed-out buffer, then it is difficult to keep recording additional tracks due to the delay.

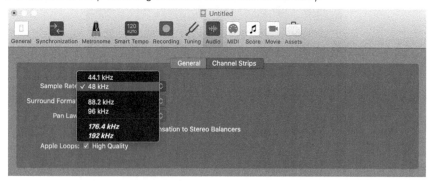

Figure 5.23 Setting the Sample Rate

Math can shed some light on the issue of the buffer and help further explain the need for resource management or finding alternate methods of controlling latency. Each project has a sample rate, which describes the number of audio samplers per second. A standard rate is 48,000 which means there are 48,000 pieces of audio data every second being used to create sound. The buffer is also measured in samples and a buffer size of 1024 (the maximum in several workstations) is equal in time to the length of 1024 samples. The formula is useful to figure out the time: 1/48,000 * 1024 = the latency induced by the buffer. That amount is approximately 21 milliseconds, which is enough to affect the performance of new recordings. If a sample rate of 41,000 is used then the latency of a 1024 buffer is even longer.

Figure 5.24 Delay Compensation Options

Delay compensation

Another factor in the latency situation comes with effects and processors which require more time to do their job and it adds delay beyond what the buffer size already creates. Nearly all sequencers have automatic delay compensation which computes the additional delay needed by a plug-in and compensates by adding the same delay to all of the other tracks in order to maintain sync between everything. The result is that there is even more latency and it is even harder to record additional instruments.

Figure 5.25 Low Latency Mode Switch

Low latency mode

Some sequencers have low latency modes which are used to help reduce latency by turning off delay inducing effects, deactivating parts of the signal chain which might cause delays, and managing the effect of the buffer. If a specific sequencer doesn't have a low latency mode then a similar effect can be achieved by freezing tracks (see the next section) or by exporting the individual tracks with latency issues, importing the resulting file, and then deactivating everything on the previous track. This helps with the latency but makes editing later more difficult.

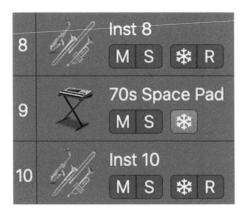

Figure 5.26 Freezing Tracks

Freezing tracks

In situations where there isn't any additional latency added through effects but the buffer is maxed out and the CPU is hitting the limit, then freezing tracks is a good option. Not only will this free up processing power but it could possibly allow the buffer size to be reduced. Freezing tracks is a feature which takes a track, records it into a new audio file with all of the effects, and deactivates all of the effects to save on processing. The key feature is that freezing the track happens without being intrusive to the workflow and doesn't add another track or even look different than before being frozen. While the MIDI can't be edited while frozen, it typically takes less than a second to unfreeze the track and then a few seconds to refreeze after the edits are complete.

Demonstration:

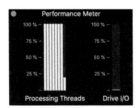

Figure 5.27

Step 1. CPU usage reaches overload

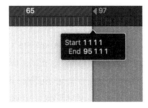

Figure 5.28

Step 2. Set project end point to prevent excess time

Figure 5.29

Step 3. Activate the Freeze option

Figure 5.30

Step 4. Select the Freeze button on the desired track

Figure 5.31

Step 5. Pick between Freeze types, in this case a full freeze

Figure 5.32

Step 6. Press Play

Figure 5.33

Step 7. Unfreeze to make a change in the future

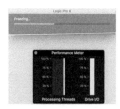

Figure 5.34

Step 8. Refreeze after the edit is complete

Recording

The recording process is critical in the workflow of orchestration. Record enable the desired track, press record in the workstation, and then play the part. If the performer is able to play the part perfectly on the first try then none of the following recording techniques are needed but they come in handy in any less than ideal situations.

Figure 5.35 Recording Example

Recording modes

In addition to recording from beginning to end, it is possible to record using a looped mode, using take folders, merge mode, using punch in/out, slowing it down, and even recording without recording. With all of these options it becomes a matter of situation and preference to decide the best way to record.

Figure 5.36 Loop Recording Settings

Looping

This is useful for repetition to get the best take possible. Looping repeats the same section over and over and constantly records whatever is played into the same track. When an acceptable recording is completed then just stop the transport.

Figure 5.37 Take Folders

Take folders

When looping, the performances can be automatically placed into alternate games or "take folders" which allow for easy editing between all of the different versions. It is possible to create the perfect take by comping (compositing) all of the best parts into a single version.

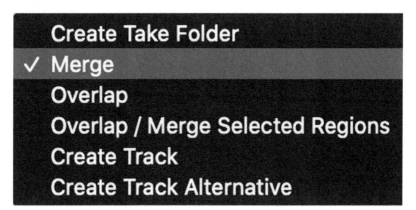

Figure 5.38 Merge Mode Settings

Merge mode

In merge mode, MIDI data which is recorded is added to the existing data. A portion of the part could be played on the first pass and then the rest on the second pass. This is

151

also a technique which is useful when recording an instrument first and then adding controller data to adjust volume, pitch, or other instrument controls. The primary benefit of using the merge function is to add multiple layers that would be difficult to add simultaneously or that would make it easier.

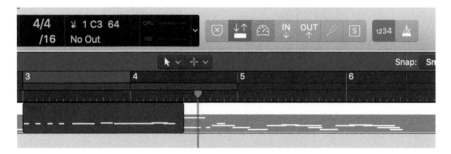

Figure 5.39 Punch Mode with Pre-roll

Punching In/Out

This option is used when a portion needs to be replaced and it can be fixed in a quick recording pass. This is an old technique used in the analog days of tape recording and was largely accomplished through manual punching by engaging the record mode in real time. In digital workstations the In and Out points can be set so that the recording engages automatically at specific points. While this is one of the options, it is used less than ever before because it is easy to simply record over a section and then edit the edges to match the replaced portion instead of setting In/Out points and focusing on making the performance so accurate.

Figure 5.40 Setting the Tempo to Half Speed

Slowing it down

This method utilizes a slowdown of tempo in order to make recording more manageable. Perhaps the performer isn't able to play the part at full speed, which is common when it comes to some string and wind parts. The important part of this option which needs to be available with a workstation is a full tempo slowdown of all tracks while maintaining the same pitch. If all tracks are MIDI then it is easy to speed up and slow down the tempo, but if there are audio tracks then it requires a different set of processing tools. Many workstations can do this, but it isn't available on all of them.

Figure 5.41 Capture Recording Mode

Recording without recording

Several workstations offer non-recording options for recording. Logic Pro has a feature called Capture Recording, which allows MIDI to be played without pressing the record button and if the part was played well, then the MIDI will be added to the track. Essentially the software is always recording the MIDI but in one case it is intentionally recorded and added to a track and in the other case it is still recorded but not necessarily added to the track afterwards. While this may not be a critical feature there are two situations in which it is useful. Perhaps in a rehearsal pass the performer plays the part perfectly but it wasn't recorded. Another situation is if a musician prefers to play the part without the pressure of recording. They can play along and then after the fact decide if a performance is worth keeping. Not every orchestrator is a seasoned performer and so it may be a feature worth exploring.

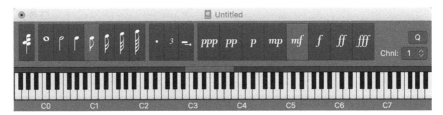

Figure 5.42 Step Recording Dialog

Step-entry

While not exactly a recording method, adding notes via step-entry is a valid form of part creation through the use of a musical keyboard and typing keyboard shortcuts. With one hand playing notes and the other typing the key associated with the desired note length, entry can be a very efficient process. The strengths associated with this method are speed without having to have piano skills, accuracy of entry, and in cases without a MIDI keyboard there are ways to still indicate the notes. The weaknesses include a lack of musical performance, it is slower than playing in the parts with piano skills, and it isn't possible to add other control data during the entry phase.

Figure 5.43 Practicing Makes Perfect

THE BIGGEST RECORDING TIP

This entire book could be written in a single sentence: write music and perform it perfectly on the first try. While it is not a prerequisite to be an excellent performer, piano skills go a long way to help with the efficiency of recording and the overall orchestration experience. The tip here is to practice performing and spend time learning the tools used in the recording process. The software workstation and the MIDI controller are the instruments and take time and effort to master; don't expect it to happen spontaneously.

Editing MIDI

Recording and editing MIDI go together in orchestration because of the nature of the musical ensemble. Instead of adding instrument after instrument until everything is there, there is a cycle of recording and editing that continues throughout the process. After a part is performed, then it is edited to make sure it fits into the orchestration. This section covers the tools used in editing to make the process more efficient and help create realistic end results.

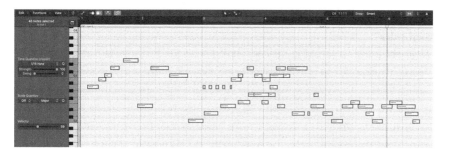

Figure 5.44 Example of Piano Roll

Basic tools

MIDI data consists of start/end times, velocity values, and various controller data. The first type of editing is moving this data around until it matches the desired performance. There will be a lot of little movements and shifting of things, and is probably the least efficient way of creating the end goal. It is possible to spend hours, days, and weeks fine tuning MIDI performances manually, but there are a few techniques which can help in the overall process.

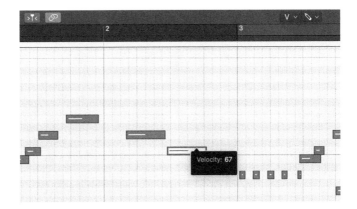

Figure 5.45 Velocity Tool

The primary location for MIDI editing is in the Piano Roll, which is called that in reference to the roll of paper used in old player pianos. Those piano rolls had rectangles cut out which are the mechanism used to get the keys of the piano to play. MIDI editing is the same process of triggering keys with rectangles, except there are a lot more things to control.

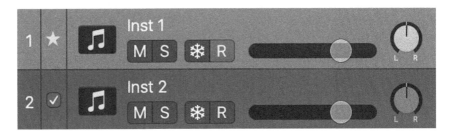

Figure 5.46 Groove Track

Groove quantization

Quantization, which aligns MIDI notes to the song grid, is used les frequently in orchestration due to the need of keeping musical parts more natural. When it is used the percentage of quantization strength is typically set to less than 100 percent in order to keep it less rigid. Instead of a traditional quantization, using a groove tool is often a much better solution. If an instrument is recorded and the timing and overall feel are right, then one way to efficiently use that timing and get other instruments to match is though the groove track. The first step is to identify the track with the desired timing, store the timing/groove, and apply the groove to one or more of the other tracks. Sometimes this process is complicated with multiple steps and other times it is as simple as a checkbox on a track.

Figure 5.47 Before and After Groove Track Activated

In Logic Pro X there is a Groove function which works seamlessly as long as the groove source is on a single track. The only way this works in an orchestral setting where parts that are ideal for the groove source are spread across numerous tracks, is if copies of each of the parts that serve as the groove source are moved into a single track and then not assigned to a sound so they won't make random noises. This is a perfect example of how far technology has come but it still eludes the power and efficiency of humans performing together.

Figure 5.48 East West Play Instrument

Instrument selection

Once the first pass of editing has finished, there is still a lot to do with each instrument. It may seem like the specific patch would have been finalized before recording but even though that would be ideal it's not always possible. A good orchestral library has a large number of instrument patches to choose between and each time a change is made then everything will have to shift. Some instruments have notes which sustain longer than others, and some have articulations which sound different than each other. If an instrument patch is switched from one to another that has a shorter length but originally it was matched perfectly with other instruments, then any number of edits will have to be made. Some notes will have to be shortened or lengthened to match the rest, along with level changes and other tweaks. It seems that having a clear vision ahead of time for the choice of instrument and style of playing means a lot of time and efforts saved later on.

Figure 5.49 Quantization Options

Quantization in all its forms can't compensate for the changing sounds of instrument patches and is only able to line up the base MIDI values. Becoming familiar with the instrument libraries is more important than almost every other aspect of MIDI sequencing and is perhaps, after the composition itself, the next most important thing overall to help with the orchestration process. In fact, there are two worlds when considering the tools used in crafting the instrumentation. While understanding the MIDI sequencing tools such as quantization, editing, delay-compensation, and all the rest is important, there are additional editing tools available in the instrument plug-ins themselves which can add or detract from the overall realism.

Figure 5.50 Logic's String Instrument

Example situation

Perhaps violin and viola parts are sequenced together using two separate MIDI instrument tracks and then later on in the process the orchestrator decides to switch the viola patch for one that has a mellower sound but has a different attack articulation. The new viola has the slightest delay as each note is triggered and a noticeable, but slight, crescendo. Adjusting MIDI in the sequencer won't easily fix this but it is possible with detail add editing to re-align all of the notes. Most orchestral instrument banks have stalked control over the various portions of the performance characteristics and it should be possible to either adjust the initial attack of the violin to match the viola or in some cases to sharpen the attack of the viola to better match the violin. There are no guarantees of a perfect match but it is important to decide between the instrument and the instrument track in terms of which is best suited to adjust the performance.

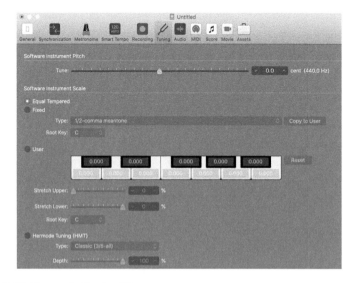

Figure 5.51 Hermode Tuning Menu

Tuning

Tuning is a significant issue in the process of digital orchestration, in part because a real orchestra is capable of maintaining appropriate tuning between the various parts of harmony. Thirds, fifths, and sevenths are all tuned differently than the exact same notes when used in other parts of the harmonic structure. MIDI, in the traditional setting, uses fixed tuning based on the tempered system which compromises all notes in order to be able to use all of the notes without adjusting tuning on a chord to chord basis. Therefore, a piano can play in all keys equally and MIDI, without adjustments, does the same thing. Most acoustic instruments are able to bend the pitch slightly in order to more accurately line up with the natural existing overtone series and so an orchestra which tunes to A-440 rarely sticks within a rigid tuning structure as instruments move around to form chords.

This is all to say that, if the goal is to have the orchestration to sound as real as possible, then tuning needs to be addressed so that the middle harmonies shift into alignment with the chords. If an orchestral project is being used as a mock-up of a piece for various involved parties to hear what it sounds like, but is subsequently recorded with a real orchestra, then it matters less about having the tuning fully locked in.

Figure 5.52 Play Instrument's Tuning Settings

Tuning systems

The ability for a workstation to adjust the tuning of instruments is widely varied and inconsistent. Pro Tools doesn't have a system in place, while Logic Pro X can tune its own instruments and some third-party instruments but not all of them. Cubase is among the best at tuning. A small minority of orchestral banks have implemented their own tuning options, but it seems like this is lower on the priority list of new releases. For example, East West's Hollywood Strings have a small built in mechanism for limited tuning but cannot work with Logic's system. They don't even reference such tuning in their manual, while Vienna Strings are both compatible and aware of the needs of tuning to create realistic reproductions.

The most commonly used method is called Hermode tuning, which is still a compromise but seems to work well enough to produce satisfactory results. This system dynamically changes the tuning depending on the MIDI content, meaning it interacts with the music and tunes notes differently at different times. The thirds and fifths are altered, and differently in major and minor chords. Thus, the workstation needs to be continuously analyzing the harmonic structure and then shifting the pitch in a variety of directions and by a variety of amounts.

Figure 5.53 Hermode Exception Checkbox

Mixed tunings

In the digital world it is possible to have some instruments play "better" in tune than in the real world. For instance, a piano, which uses the Equal Tempered system as a compromise to play in all twelve keys, can now be played using the Hermode system. It sounds different and wrong at first since the dissonance caused by everything playing shifted in the tempered system is something that is now normal. Of the three possibilities of tuning orchestra and piano, it sounds acceptable for the piano to remain tuned as is traditional while the orchestra is in Hermode tuning and it is usual for both to be tuned using the Hermode option. Tuning both in the Equal Tempered system leaves a lot to be desired, with a dull and potentially lifeless result.

Options for no tuning system

It is possible to tune individual notes using the pitch bend data, but it can get complicated fairly quickly. This requires the specific notes to be on their own instrument track and therefore in their own instrument plug-in. MIDI 2.0 allows individual notes to be pitch bent but most tools haven't fully adopted this standard yet. If the goal is to make an orchestration which translates directly into a printed score, then this technique certainly blocks that possibility

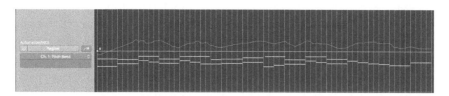

Figure 5.54 Automate Tuning Data

Another option is to automate the tuning system of the orchestral instrument plug-in. This solution varies depending on the capabilities of the instrument but it is highly likely to have some sort of tuning option that can be automated. If a specific instrument is critical to the orchestration, but the tuning remains an issue, then consider hiring a single musician to play the part and create an audio version of the part played perfectly in tune by a human. The last thing to consider is that if tuning is a critical function of the orchestrations at hand, then use a workstation which is capable of doing what is needed.

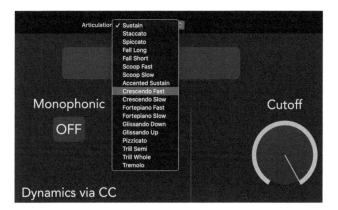

Figure 5.55 Articulation ID System

Articulations and styles

As described earlier, virtual instruments are able to recreate instruments accurately via multiple methods. If a trombone part builds over time, then there are several ways to accomplish it. A patch could be used which is a recording of a trombone player performing a crescendo over time. This exists, although they are limited in their usability due to the fact that the original recording has a set length and can't be adjusted easily, or at all, in the sequencing process. It might work to shorten it a little, but it is certainly impossible to make it longer. Another method is to use a patch which has good sustain but no existing build, and to use a controller to have the volume increase as the note sustains. Some instruments are designed to imitate the real instrument and as dynamics are controlled the sound doesn't just get louder but it also increases in intensity and mimics the timbre change that happens when a brass instrument, for example, is played louder. If the instrument being used doesn't accurately recreate the sound, then it's possible to add volume and then use an equalizer to brighten the sound simultaneously. It isn't a perfect replacement but better than just increasing the volume

Figure 5.56 KeySwitching Example in Play

Patch switching

There are a few ways that are used to switch between articulations and styles, which can be categorized by whether they are optimal for performance or not. Keyswitching and mod-wheel changing are useful when performing and things like score to MIDI functions and articulation IDs are less designed for live performance. All methods are just different ways to switch between different parts of the instruments. A keyswitch uses MIDI notes lower or higher on the keyboard to switch between sounds. The same switching process could be attached to the mod-wheel, drum pads, or anything which has MIDI output. This group is efficient at switching as long as the trigger is within reach of your hands or feet when performing. The other type of switching is tied into the sequencer with editing required. Several workstations use markings in the score editor to switch between instrument styles. For instance, if in the score a note has a staccato punctuation then the instrument will play a staccato patch. This is extremely useful for those who are creating orchestrations in the score editor instead of the piano roll editor. Other workstations attach articulation/style information directly to each MIDI note, which makes it easy to change things on a note by note basis.

Notation

MIDI is useful for controlling instruments and creating complex orchestrations but not something that musicians can read when playing real instruments. This section is perhaps unexpected because instead of covering the basics of notation and the associated tools, the following paragraphs instead take a closer look at the relationship between MIDI and the score. There are specific ways that the score functionality of most workstations supports the orchestration process.

Figure 5.57 Notation Toolkit

The purpose of the score

Score editors which are attached to MIDI sequencers are uniquely positioned to do much more than applications such as Finale or Sibelius, which in fact are very powerful notation editors in their own right. If the end goal is to have a printed score that can be used in a performance or recording session, then the score is orchestra centric and means the entire project would be laid out to match an orchestra. If the end goal is to create a realistic mockup or realistic performance, then the number of tracks and instruments don't matter and it doesn't matter what the score looks like. The way the score is used in this situation can be wildly different, unconventional, and just plain wrong.

Figure 5.58 Visual Quantization in the Score

Visual quantization

It is common to have separate quantization settings for the piano roll and the score editor because something that works musically might not work in the score if shown with exact timing. For instance, if a score is put in front of a musician and they play the notes there is an expectation for the music to be interpreted, which includes liberties taken in time. Well, recording the performance first with all of the timing variations and the ebb and flow make it difficult for an accurate score to be created unless separately quantized. Even then there are reasons to turn off the quantization so that the notation is an accurate representation of the MIDI and then it can more directly be used to edit the MIDI. Perhaps it is best to find a good balance in adjusting the visual quantization settings so that it makes sense to look at and when editing.

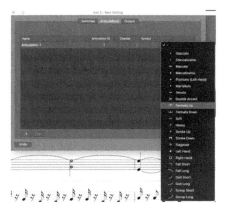

Figure 5.59 Articulations are Tied to the MIDI Data

Performance modifications

The most powerful notation editors tie the MIDI and score together in a two-way relationship. This means that the MIDI notes create the score but if a crescendo is added in the score or other notation specific markings such as staccato, accents, or other score things then these actually adjust the MIDI data and the performance. While perhaps this makes sense on the surface, not all notation editors offer this tight integration. Pro Tools has nothing like it. Logic Pro has a simple version of it. Cubase has the best score integration and is one of the top choices for composers and orchestrators.

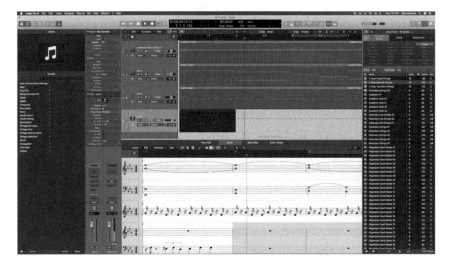

Figure 5.60 Recording with the Score Visible

Notation in recording

Perhaps it would make more sense to have written about the score's use in the recording process above in the section on recording but it is here to keep all of the score elements together. One of the biggest limiting factors of recording MIDI is that it all just looks like rectangles and, if a musician can read music, then it makes sense to integrate the score into the process. After recording a part, then perhaps make the score visible and, as the next part is recorded, the existing parts are visible. It helps with getting the notes right and focusing on the orchestration.

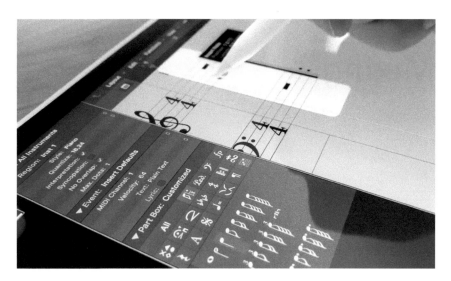

Figure 5.61 Handwriting Options in Notation App

In addition to recording with the score being visible, there are also a few notation editors which accept handwritten parts that are subsequently converted to MIDI. No keyboard, no record button, and only a stylus on a screen. It is an interesting return to "old-school" techniques using cutting edge technology. While perhaps performing during the recording process gets some orchestrators to the final product in a much more efficient manner, some might prefer to draw in the parts if the technology provides a solid path towards the final production. Perhaps the holy grail of orchestration is a notation workstation on an iPad with the Apple Pencil. Having a multi-touch interface in a paper sized handheld device is potentially the most powerful way to work. Add in the ability for the app to interpret handwritten entry into MIDI and that's the full kit and caboodle. This is not a dream and already exists, which is a testament to the advancement of modern technology.

Summary

The tools in this chapter represent the most commonly used when creating digital orchestrations, and certainly there are specific tools which each workstation offers to enhance the production. The most important piece of advice is to always keep the music in the forefront of the process, since that will help guide every decision and keep the technology appropriately harnessed.

Chapter 6

Mixing

Mixing is one of the final stages in a production and often times is the most important. Mixing is the stage where all instruments are balanced together to create the complete listening experience. With a properly mixed piece, the listener can get lost in the music and become engulfed in the piece as a whole. When a project is mixed well, nothing sticks out unintentionally; the timpani do not cover up the woodwinds, the clarinets are not buried under the horns and so on and so forth. Instead, the music is heard as a sum of all its parts, and the orchestra becomes an instrument in and of itself. Despite hearing the orchestra as a single sound, the listener should be able to pick out each individual instrument or instrument group with careful listening. This balance of creating a cohesive sound while maintaining instrument individuality is an art form and is the reason why good recording and mixing engineers are held in such high esteem.

Figure 6.1 An Orchestra in Rehearsal

Before we dive into mixing in depth, it is important to understand how mixes were achieved throughout history in order to gain a better understanding of the craft. The very first recordings were done on wax cylinder machines like the Edison Phonograph and the Gramophone. These devices contained a sound capturing horn which the musicians would huddle around and play into. Sound traveling down the horn would then excite a small diaphragm at the smallest part of the horn. The diaphragm was connected to a carving needle that sat atop a rotating wax cylinder. Sound exciting the diaphragm would cause the needle to move and in turn carve grooves into the moving wax cylinder beneath. Upon playback, this process would be reversed: The needle would be dragged along the newly cut grooves which would excite the diaphragm causing sound to emanate from the horn

Figure 6.2 The Edison Recorder

Early recordings mainly featured spoken word or a single instrument. But as these novel recording and playback devices increased in popularity, demand soared for popular music played by a group of musicians. Placing multiple musicians in front of the sound capturing horn however, proved to be troublesome. For example, if you were to have a drummer, pianist, and vocalist, it would be very easy for the drums and piano to drown out the vocalist during louder passages. Engineers quickly realized that in order to create a mix where the listener would be able to hear every sound, musicians would have to stand at varying distances from the sound capturing horn. In the previous example, the drummer would be placed further from the sound capturing horn than the piano and vocalist. This allowed all instruments to have a balanced loudness. This form of physical distance mixing was not a new concept. As we will learn later on, this approach has been used in orchestras for centuries. However, this was the first instance of intentional mixing for recorded music.

Mixing practices

As recording technology advanced, so did mixing practices. The next big advancement was the introduction of amplification. Wax cylinder recorders required relatively loud sound so that the grooves could be cut cleanly in the wax. This often resulted in musicians and vocalists having to play or sing louder than they may naturally in order to obtain a quality recording. Microphones and amplifiers changed all that. Now musicians could be placed around a single microphone, which then fed an electrically driven cutting needle. Although musicians still had to be placed at varying distances away from the microphone to achieve proper balance, the overall sound level could be lower at the microphone because the engineer could amplify the resulting signal before it reached the cutting needle. This allowed for better musical performances. Recording studios were also evolving at this time and were starting to have better acoustics in order to achieve better sound.

Tape machines

The next recording evolution came in the aftermath of World War II. American military officers discovered that the Germans had created a fairly sophisticated way of recording sound onto magnetic tape. The American military brought this technology back, and it quickly began being put to use in recording studios. Originally, magnetic tape recording only allowed for a single sound source (one microphone). However, this single sound source recording was quickly replaced by the invention of multi-track recording.

Figure 6.3 24 Track Tape Machine and Transport

Essentially, multi-track recording allows for multiple sound sources to be recorded onto the tape separately. This means that instead of setting up the musicians around a single microphone, a single microphone can be placed in front of each musician. This allows the engineer to individually mix the microphones together to create a good balance.

Multi-track recording also allowed for musicians to record their respective parts separately at different times. This greatly increased sound quality because now a single musician could re-record their part if they made a mistake rather than the whole group having to re-record. It also allowed for completely isolated recordings, which resulted in a higher fidelity mix

Another major advancement that magnetic tape recording brought to the mixing field was the ability to mix the music after it had already been recorded. Wax cylinder and early metal disk recording would physically carve out grooves in the playback medium (cylinder or disk) as the music was being played. This meant every sound that entered the sound capturing horn or microphone was recorded. There was no way to mix the levels after the sound was imprinted onto the medium. Magnetic tape recording changed this. In tape recording, audio signal is stored on the tape before getting transferred to its corresponding playback medium (in most cases this would be vinyl records). By having

Figure 6.4 Mixing Console

Figure 6.5 Digital Audio Workstation

the recording stored on the tape prior to getting pressed into vinyl, engineers were free to play the music back as many times as they wished while mixing the levels of the various microphones.

Having the ability to adjust the individual sounds after recording was a crucial evolution in mixing. Rather than solely being able to adjust volumes, engineers were now beginning to experiment with changing sonic characteristics of the audio. This spawned an entire industry of companies producing equipment designed with the distinct purpose to change various sonic elements of recorded audio. Currently, mixing makes extensive use of this equipment, known as signal processors, which will be examined in greater detail later in this chapter.

Mixing and recording technology evolved throughout the 1970s and 1980s. Advancements included higher fidelity equipment, better knowledge and execution of studio room acoustics, increased tracks available to record on, etc. However, the next large zeitgeist in mixing and recording technology came with the advent of the personal computer and digital recording. With digital recording came the ability to visually see waveforms, a crucial element when performing precise edits. This again changed how engineers mixed because they now had the ability to manipulate and edit minor in-consistencies in audio

All of these advancements in mixing have built off of each other rather than replace each other. A good mix still relies on properly placed musicians, properly placed microphones, high fidelity equipment, and skillful manipulation of the audio in the vo-lume, sonic, and physical domains. When all of these elements are put together, the result is a well-balanced mix.

Mixing as an art form

Now that we have a basic understanding of how mixing has evolved through recording history, let's take a closer look at mixing as an art form. Modern mixing can be broken down and examined in three dimensions: loudness, spectral, and spatial. All sounds and

Figure 6.6 Loudness Meters

instruments must be balanced in these three domains to create a well-balanced mix. How the sounds and instruments are balanced may vary and is up to the mixing engineer. In a properly balanced mix, each instrument has its own space and is not competing with other instruments. Engineers can manipulate each sound's loudness, spectral, and spatial characteristics to balance each instrument. Here is an examination of these three dimensions to obtain a better understanding of them.

Loudness

Loudness is perhaps the most self-explanatory of the three dimensions – it refers to how loud the sound/instrument is in the final mix. When two sounds are played together, and one cannot be heard, the engineer must simply increase the volume on the quieter sound. However, special care must be taken to keep the mix as a whole balanced. The mixing engineer must always keep in mind various instrumental relationships to one another in order to create a realistic loudness balance. For example, it would not make sense for a tambourine to be louder than a bass drum. Likewise, a flute should not overpower an entire brass section. Increasing the volume of softer instruments while maintaining their proper orchestral balance is key

There is a loudness ceiling on all audio recording and playback equipment. Sounds that surpass this loudness ceiling will "clip" and become distorted. Therefore, the engineer must never allow sounds to surpass this level while mixing. Sound is additive; the more sounds that are combined, the louder the overall output will be. This often proves difficult in mixing because although no individual sound might be surpassing the output ceiling, the mix as a whole can be. In certain instances, one instrument may be a little too soft, and, when bringing this instrument up in volume, it pushes the entire mix over the output ceiling and introduces clipping. Because of these instances, it is usually recommended to decrease other sounds in loudness around the quieter instrument. This may seem counter intuitive, but it gives the mixing engineer more room to adjust individual loudness later on.

Figure 6.7 Faders in Logic Pro X

Spectral

The next characteristic we will cover is spectral. Here we are referring to the sonic spectrum, or frequency range, of the mix as a whole, as well as the individual sounds and instruments. Professional equipment is generally designed with a range of 20 Hz to 20 kHz in order to cover the typical range of human hearing, although it is widely recognized that some people have a larger range and others have a narrower range. (Hertz, or Hz, is a unit of measurement relating to cycles per second, or how many times in a second a waveform completes its full cycle.) In order to create a well-balanced mix in regard to the sonic spectrum, the overall frequency range of the mix must span the entire human hearing range

A mix that is lacking in the low frequency range will sound thin and be lacking "bass," where as a mix lacking in high frequency will sound muffled and "dark." In terms of mixing, engineers typically divide the human hearing range into various frequency bands. These bands are represented as: Low, Low-Mid, Mid, High-Mid, and High. With the frequency range divided into equal bands, the engineer must sort which instruments fit into which frequency band. A balanced mix will not only cover the

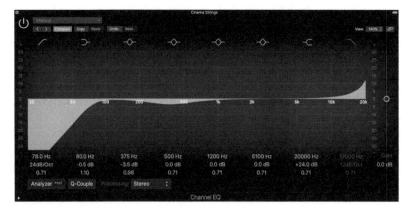

Figure 6.8 Equalizer

Figure 6.9 Surround Pan

entirety of the human hearing range, but each frequency band will be represented equally. Although this balancing is done in mixing, it begins in both the composing and recording stages of a project. Making sure the instrumentation of a project spans the various frequency bands is crucial for ending up with a well-balanced sonic mix.

Spatial

The final dimension in mixing is the spatial dimension, which is where a sound is heard by the listener. In modern stereo and multi-channel recordings, this allows the engineer to place sounds directly in front of a listener or to either side of them, as well as closer and further from them. The ability to move sounds to different areas opens up more potential ways for the engineer to balance various sounds together. For example, if there are two sounds that need to be around the same volume and take up a similar frequency range, the engineer can place one of the sounds slightly to the left and one sound slightly to the right in order to spread the sounds out more and allow room for each around to exist. However, the spatial aspect of mixing is not only limited to left and right; engineers can also place sounds closer or further from the listener. Later in this chapter 360° audio is also explored which brings the listener into a fully immersive experience

In the early days of recording when louder instruments were placed further back in the room, the instruments sounded further away. As recording advanced, it was no longer necessary to place louder instruments further away. Instead, the desire to artificially place them further away arose. This is because engineers can make room for various sounds by placing them at different distances from the listener. This is the same principle as making room by moving sounds to the left or to the right.

Much like with the loudness spectrum, engineers take certain precautions when mixing in the spatial domain. For example, it would not make sense to have the bass drum and cymbals right in front of the listener while the timpani and vibraphones are farther back. Engineers will typically employ a realistic vision of where musicians are placed in a performance when making spatial changes in the mix.

Figure 6.10 Luggage for the Mix

By combining these three mixing spectrums, engineers have a wide variety of room to make sure every instrument is heard appropriately. I like to compare mixing to fitting luggage in a car. You have a certain amount of luggage and a certain confined space it needs to fit into. You can move pieces of luggage further back, stack luggage on top of each other, or move pieces of luggage to the left or to the right. The same thing is true with mixing. There is a confined space in the amplitude and stereo domains, and the engineer must fit all of the instruments and sounds in that space while making sure every sound can be heard. By manipulating the sounds within these three spectrums, the engineer can create a well-balanced mix

Much like in the early days of recording and mixing, orchestras have employed a physical mixing approach for creating their mixes for the listeners. At a performance, the various instrument groups are arranged on the stage in a way so that when the emanating sound reaches the audience, it has blended appropriately into a well-balanced mix.

It is important to examine the various instrument groups of the orchestra and study where they are typically placed on the stage for optimal sonic balancing. Strings are typically placed closest to the conductor with a left to right spread from violins to violas, and then cellos. Basses are typically placed a little further back and on the right. Woodwinds are usually placed behind the strings in two sections. The first, closest section will include flutes, piccolo, oboes, and English horns, while the second section (behind the first) will include clarinets, bass clarinets, bassoons, and contra bassoons. Immediately behind the woodwind section, there is the brass section spread out in the back of the orchestra. Harps, pianos, and percussion will typically be towards the back on the left.

Figure 6.11 Orchestra Chart

While louder instrument groups such as percussion and brass might typically overshadow softer instrument groups, this positioning places these louder groups further back, resulting in more sound dissipation by the time it reaches the audience. This positioning allows the various instruments to perfectly blend together when their sound reaches the audience. The listener hears the orchestra as a whole rather than individual instrument groups. Since the orchestra is also spread across the stage, the sound is also able to envelope the listener in the stereo spectrum

While this physical mixing works for live performances, it is not wholly sufficient in the modern recording studio. Engineers will typically place microphones on each instrument for better fidelity. Because of this, level mixing will almost certainly have to be performed. This individual miking also eliminates natural reverberation that is heard at an orchestra performance. Therefore, engineers will typically add artificial reverberation in the mixing process. Lastly, microphones do not pick up sound the same way our ears pickup sound. Therefore, some sonic manipulation will typically be performed to ensure the recording sounds good to the listener.

Figure 6.12 Pit Orchestra Example

Figure 6.13 Post Production Mixing Console

As stated earlier, once multi-track recording gained prominence and mixing became a celebrated art form, an entire industry of sound manipulation devices became available for aid in mixing. These devices are known as signal processors, and there are countless varieties on the market. The vast majority of these signal processors can be separated into three distinct categories: dynamic based signal processing, spectral based signal processing, and time-based signal processing.

Dynamics

Dynamic based signal processors are devices which alter the dynamic range of a sound (e.g., the difference between a sound's softest and loudest parts). The most common devices in this family are compressors, limiters, expanders, and gates. Each of these devices offers unique features that can help an engineer either reign in, expand on, or otherwise alter a sound's dynamic range

Figure 6.14 Compressor

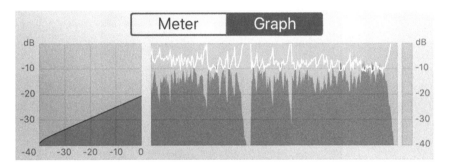

Figure 6.15 Compression Reduction Graph

Compressors

Compressors are the most commonly used dynamic based signal processor. As their name suggests, compressors work by diminishing, or compressing, the overall dynamic range of a signal. As a signal enters the compressor, the peaks of the signal are lowered in amplitude at a rate and amount determined by the user. The signal can then be amplified resulting in a signal with a smaller dynamic range. When used effectively, compressors can bring forward subtle elements of an instrument that are otherwise overshadowed and lost in a mix. A compressor can also add fullness to a sound that might otherwise be weak and thin. Although there are hundreds if not thousands of different compressors on the market, most will have similar settings and controls for the engineer to manipulate

The first control we will examine is threshold. The threshold is the setting which dictates how loud a signal must be in order for the compressor to start working. The user simply sets the threshold and any sound which is lower than threshold will be left alone, while any sound which surpasses it will be reduced in amplitude. The lower the threshold, the more often the compressor will be working. If the threshold is too high, the compressor may never actually be compressing the signal

The next control we will examine is the ratio. This control works in conjunction with the threshold and determines how much the compressor will lower the amplitude of

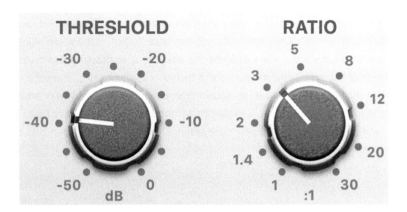

Figure 6.16 Ratio and Threshold

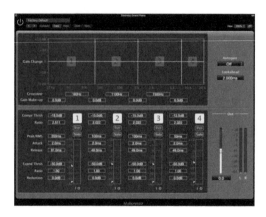

Figure 6.17 Alternate Compressor

a signal once the threshold is surpassed. Therefore, the ratio determines how much compression will be added to a signal. While the threshold is typically displayed as decibels (dB), the ratio is displayed as a true ratio (i.e., 1:1, 2:1, 3:1, etc.). These numbers represent how many dBs a signal will be attenuated by in relation to how many dBs it is over the threshold. For example, a 2:1 ratio would mean that an input signal of 2 dB above the threshold will be attenuated down to 1 db above the threshold. Likewise, a signal that is 8 dB above the threshold will be attenuated down to 4 dB above the threshold. Therefore, the ratio directly relates to how much a signal will be attenuated by

After setting the attack and release of a compressor, the mixing engineer will want to control how fast the compressor attenuates a signal, as well as how fast the compressor returns back to a normal non-compressing state. This is accomplished with the attack and release controls respectively. Both the attack and release controls are typically displayed as increments of time usually ranging from milliseconds to seconds. There are no default attack and release settings. Instead, they will change depending on the unique aspect of the sound entering the compressor. For example, it may seem logical to set the attack as fast as it can go, causing the compressor to begin attenuating the instant the threshold is surpassed. In reality, this would be detrimental to a highly transient and percussive sound, such as a snare drum. With an instantaneous attack setting, the initial "snap" of the snare drum gets attenuated and squashed causing a muffled sound. With a slightly slower attack, the initial "snap" of the snare triggers the compressor, but the actual attenuation begins after the snap has subsided causing the ringing out of the drum to be attenuated and then eventually brought up in volume. This creates a much thicker and beefier sounding snare drum. Because every sound will have a unique characteristic, it is important for the engineer to critically listen to the sound and determine which attack release settings will work best for a particular sound.

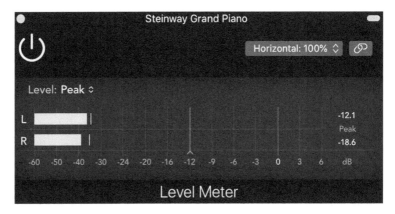

Figure 6.18 Use a Meter to Gauge Compression

The final compression setting we will examine is make-up gain. Since a compressor works by attenuating signals, it would make sense that the signal at the output stage of the compressor will be quieter than the signal at the input stage. Therefore, all compressors will feature some type of gain control which allows the engineer to amplify the signal at the output stage. Most compressors will also feature a gain reduction meter which displays in real time how much gain is being attenuated from a signal. When setting the makeup gain, all the engineer needs to do is increase the makeup gain by the same amount that the signal is being attenuated in order to create transparent compression.

By artfully adjusting each of these settings, instruments can now sit in the mix when they would have otherwise been buried at various louder sections. A well-placed compressor will also "level out" a performance, meaning that the overall loudness of the instrument throughout the performance will stay consistent. Although compressors were designed to be transparent and used delicately, by cranking the various settings, the compressor becomes more of an effect itself and can be used artistically in various manners

At the opposite end of the dynamic based signal processor range, is the expander. Unlike a compressor, an expander increases a sound's dynamic range. An expander will artificially make the quieter sections of audio quieter while increasing the

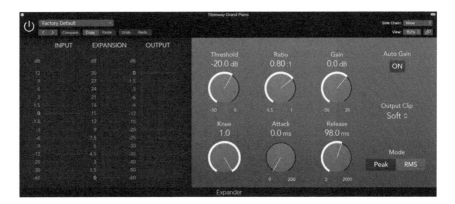

Figure 6.19 Expander

Figure 6.20 Gate Processor

louder sections. Expanders tend to work very well when used in conjunction with compressors for hyper-compressed artistic effect.

Gates

The next dynamics-based signal processor we will examine is a gate. These devices prevent unwanted sounds such as bleed from other instruments, buzzing/humming, and background ambient noises from being heard. Like on a compressor, the user sets a threshold. Any sound quieter than the threshold level is prevented from being heard. Sound that is louder than the threshold is passed through

If a gate were to instantaneously mute and un-mute audio, the resulting sound would be choppy and not particularly useful. Fortunately, all gates offer the user the ability to set attack and release controls which dictate how quickly the gate mutes sound once it drops under the threshold and how long it takes to fade the sound back in once the threshold is surpassed. These controls greatly reduce the choppiness caused from the muting and un-muting. Most gates will also feature a control called hold. The hold control allows sound to pass through the gate for a user determined amount of time after the threshold is surpassed. This is crucial for a natural sound. If a gate was placed on a timpani, for example, the gate would open up when it is first struck, but it would close very quickly because the "ringing out" of the timpani is much quieter than the initial transient caused from the mallet strike. By increasing the hold control, the gate will stay open and allow for the entirety of the timpani signal to be heard.

Limiters

The last dynamics-based signal processor we will discuss are limiters. At its core, a limiter is basically just a compressor with an extreme ratio setting. A limiter will completely prevent a sound from reaching a user set loudness level. The limiter will do whatever it needs to squash the sound in order to prevent it from reaching the output ceiling. Limiters are very useful in both recording and live settings in order to prevent a sound from clipping and distorting. Limiters are most often used in the mastering phase of a recording production in order to increase the overall loudness of a production.

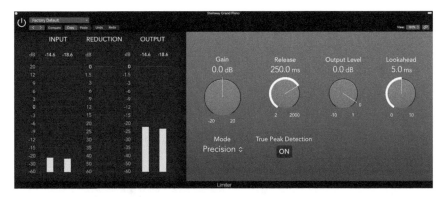

Figure 6.21 Limiter Processor

When mixing for orchestra, many of these dynamic choices will be dictated and executed through a combination of the composition and the performer. The conductor will typically convey real time dynamic direction to the musicians which will be paired with the dynamic direction written on the sheet music. Dynamic based signal processing will be used more for potential shortcomings and limitations in the recording itself. Dynamic based signal processing may not even be utilized until the mastering phase of an orchestral recording. In contrast, it would not be uncommon to see a compressor on 75 percent of the tracks in a modern pop recording.

Spectral processing

The next processing family we will examine is spectral based signal processing. These signal processors alter the overall frequency range of a sound. As stated earlier, a well-balanced mix will sonically cover the entire human hearing range from extreme low frequency up to extreme high frequency. Each frequency band should be heard equally in a well-mixed project. The listener should also be able to hear each instrument clearly. This often proves difficult because many instruments have frequency ranges that overlap with each other. A skilled mixing engineer can use spectral based signal processors to carve out sonic space in order to make every instrument and sound fit perfectly in the mix.

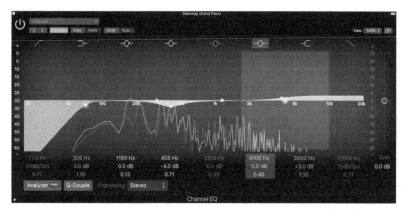

Figure 6.22 Typical EQ Curve with Active Analyzer

The most common type of spectral based signal processor is known as an equalizer, or EQ for short. An EQ allows the user to boost or cut specific frequencies of a sound. EQs are extremely useful for not only carving out sonic space for each instrument, but also for fixing or even enhancing the sound of an instrument. Just like with any other signal processor, there are thousands of different EQs available that each sound slightly different. At their core, all EQs are comprised of a number of frequency filters of varying shapes that can be manipulated. It is important to understand these various filter shapes in order to know how an EQ will alter a sound

The first type of filters we will discuss are known as pass filters (also called cut filters). There are three different options: low pass, high pass, and band pass. A low pass, or high cut, filter cuts off all frequencies above a given frequency while allowing lower frequencies to "pass" through. This filter shape is commonly found on synthesizers and can be used to instantly get rid of high frequency on a sound. A high pass, or low cut, filter does the exact opposite. High pass filters cut out all frequencies below a given frequency while allowing higher frequencies to pass. These filters can be used to get rid of low frequency rumble and low ambient noise. The final pass filter is known as a band pass filter. These filters basically combine both a high pass and low pass filter which cuts out all frequencies above and below a given frequency while allowing that

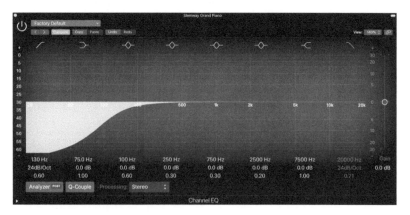

Figure 6.23 High Pass Filter

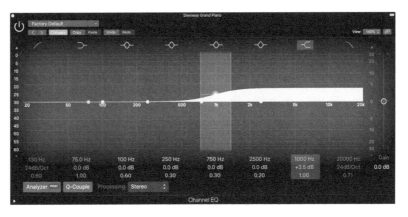

Figure 6.24 Shelf Filter

frequency band to pass. This filter is used most commonly to create the classic "telephone" effect which causes the audio to sound like its coming through a small telephone or bull horn

Another common filter shape is the shelving filter. Like pass filters, shelving filters affect all frequencies above or below a given frequency. Unlike pass filters, shelving filters allow the user to cut the frequencies by any amount rather than completely cutting them off. Shelving filters also allow the user to boost the frequencies by any amount. They are found in two varieties: high shelf and low shelf. As their names suggest, a high shelving filter effects all frequencies above the given frequency, while a low shelving filter effects all frequencies below the given frequency range. These filters can be used to boost or cut high and low frequencies, allowing the engineer to slightly add or remove brightness or low end easily. Shelving filters can be boosted or cut at varying degrees ranging from subtle to extreme allowing for a lot of sonic changes

The final filter shape is the notch filter. Unlike pass and shelving filters, notch filters only effect a single frequency. The user is able to boost or cut the selected frequency by as much or as little as needed. Most notch filters will also feature a variable bandwidth control, or "Q," that dictates how wide the frequency band being

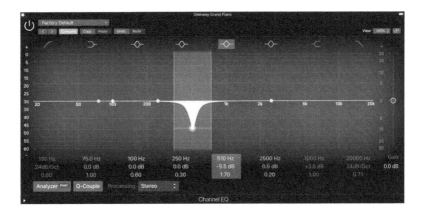

Figure 6.25 Notch Filter

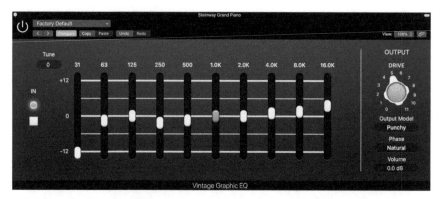

Figure 6.26 Graphic Equalizer

adjusted is. A narrow Q would result in just the chosen frequency being boosted or cut while a wide Q would boost or cut a wide array of frequencies both above and below the chosen frequency. Notch filters are useful for lowering unwanted resonant frequencies on a recording

As stated earlier, an equalizer is a signal processor that is comprised of a number of these filters. An engineer can use these equalizers to slightly or drastically change the frequency content of a given sound. EQs typically come in two distinct variations: graphic and parametric. A graphic equalizer features a number of individual notch filters spaced across the human hearing range. Each notch filter will have an equal amount of space to be boosted or cut. The engineer can adjust a single frequency band or an entire range allowing them to easily create any of the aforementioned filter shapes. Parametric equalizers also cover the entire human hearing range but typically with less controls. A parametric equalizer will typically feature a low frequency control (usually a shelving filter), two or more middle frequency controls (notch filters), and a high frequency control (usually a shelving filter). With these four controls, the engineer is able to alter the entire sonic spectrum of a sound or instrument. Like most signal processors, equalizers can be standalone units, digital plug ins, or can be found on multi-effects units. Unlike the rest of the signal processors, EQs are often included on every channel of a mixing console. This is because it is not uncommon for an engineer to adjust the EQ curve on most tracks being mixed. This is especially true in modern pop recordings.

Figure 6.27 Mixer Example

Time processing

The final family of signal processors we will cover is time-based signal processing. As the name suggests, this family of signal processors includes any effect which alters a sound in time. The most common effect in this family is reverb. A reverb unit attempts to mimic the sound reflections that are present when playing an instrument in a highly reflective and reverberant room or concert hall. A reverb unit accomplishes this by playing slightly delayed versions of the sound at varying rates and levels

Reverb

Engineers began experimenting with adding artificial reverb to recordings in the late 1950s and early 1960s. The very first iterations of artificial reverberation came in the form of echo chambers. Typically built inside recording studios, echo chambers were relatively small, highly reverberant rooms. Inside the room, there would be a loud speaker and a microphone. The engineer would pipe music into the room via the speaker and would then record the resulting reverberated sound through the

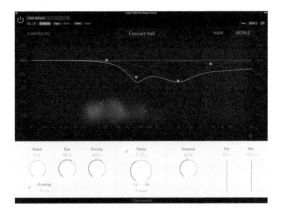

Figure 6.28 ChromaVerb

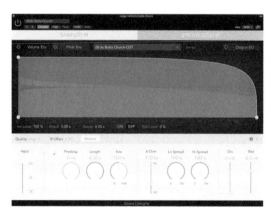

Figure 6.29 Another Reverb

microphone. The engineer could then mix this new "wet" (otherwise known as ef-fected) version the non-reverberated "dry" signal.

As recording technology evolved, so too did artificial reverb. Echo chambers began being replaced by large plate reverbs (large pieces of sheet metal that were excited into vibration via a speaker driver and the subsequent sound captured by a pickup). Plate reverbs were replaced by smaller spring reverbs (same principal as plate reverbs but the sheet metal is replaced by a suspended spring). Finally, physical reverbs were mostly replaced with digital reverbs

By adding artificial reverberation to an instrument, the engineer is able to make the instrument sound like it is extremely close or extremely far away and ev-erywhere in between. This is an invaluable tool in mixing. As stated earlier, mixing is a balancing game. If two sounds need to be heard at the same time and take up a similar frequency range, the only option is to move them spatially. One option would be to pan them away from each other in the stereo field. Should both sounds be main melodies that need to be in the center, the only other choice an engineer has is to spread them out from front to back with a reverb

The use of artificial reverberation is not limited to commercial pop music.

Figure 6.30 360° Reverb

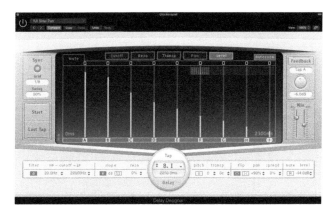

Figure 6.31 Delay Designer

Artificial reverb plays an immensely important role in the mixing of orchestral recordings as well. Listeners are used to hearing orchestral instruments naturally reverberate in concert halls. This natural reverberation is arguably part of the timbre of these instruments. When recording in a studio, much of this reverberation is absent. The resulting recording will sound somewhat lacking due to this absence of reverberation. Therefore, it becomes essential to add reverb during the mixing phase. This is especially true when attempting to blend synthesized orchestral instruments with real orchestral instruments. A synthesized string or horn sound will significantly stand out in an orchestral recording without heavy processing to blend it.

Delay

Although reverbs are the most common of the time-based signal processors, there are others available to mixing engineers. The next most common time-based signal processor is a delay. Like a reverb, a delay plays slightly delayed versions of the sound However, a delay will typically be programmed so that the delayed versions are distinctly heard after the initial sound. When used subtly, a delay can add depth to a sound much like a reverb. When used heavily, a delay can almost become a crucial part of the sound adding to the rhythm and melody. By altering the delayed sound in various ways, a number of effects are created, such as chorusing, flanging, and phasing. These effects can be used to further alter the sound in the attempt of balancing all instruments together.

Mixing process

Now that we have a thorough understanding of the what is needed to create a well-balanced mix as well as the various mixing tools engineers have available to them, let's put it all together. As stated many times in this chapter, a good mix is balanced in all domains (loudness, spectral, and spatial). Therefore, a good mix starts with the song composition and instrumentation choice. A balanced composition will have parts written

Figure 6.32 Microphone Placement

for instruments spanning the entire frequency range. The next step in creating a balanced mix takes place during the tracking phase. Adequate microphone choice and placements should be selected to ensure that the engineer is able to capture not only the unique characteristics of the various instruments, but also emphasize each instrument's unique placement in the frequency range

During both the tracking sessions and the mixing sessions, the engineer can carve out sonic space for each sound using equalizers. It is helpful to begin by removing frequencies that are not necessary. For instance, a piccolo does not have frequency content in the low frequency range. The microphone however, will still be picking up low frequencies either from ambient room noise or bleed from other instruments. Therefore, placing a high pass filter on the piccolo track and cutting out low frequencies will help clear space for the instruments that have a wealth of low frequency content like basses, tubas, timpani and bass drums. The engineer can also use equalizers to forcibly carve space out for various instruments. For example, basses and bass drums often times fight for sonic space due to having similar frequency ranges. One possibility for creating space for each would be to EQ each instrument in an opposite manner, such as cutting 200 Hz on one while boosting it on the other.

While crafting the overall sonic spectrum, the engineer will be simultaneously balancing the loudness spectrum. Again, a well-balanced loudness spectrum starts with composition and performance. During the mixing session, the engineer can balance each sound's loudness even further. By using dynamic based signal processing in combination with adjusting the instrument's fader levels, the engineer will be able to create a dynamic and well-balanced loudness spectrum. It is important to note that most mixes will require some sort of level automation (pre-programmed adjustments set up in the mixing software). Due to dynamic changes of various passages as well as performance inconsistencies, instrument levels will rarely be set at one level and remain there for the entirety of the song. Most modern digital audio workstations (DAW) will feature extensive automation options for programming these changes in sequence with the project.

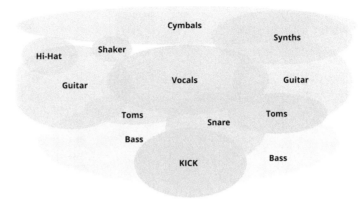

Figure 6.33 Visual Mix

With a well-balanced dynamic and spectral field, the engineer is left with balancing out the spatial field of the recording. This will typically follow a similar layout to how an orchestra is setup on a stage. Engineers will use the pan control to dictate where a sound is in the stereo field, and they will use reverb to dictate where they are from front to back. Although the spacing of instruments is typically related to their physical placement in a concert hall, some instrument or instrument groups might be moved slightly or exaggerated in space to make room for each sound to be heard clearly.

The mix can then be bounced down, or rendered down, from its multitrack format to its release format, whether that is a stereo audio file or a surround audio file. This rendered mix can then be sent off to the mastering phase where some additional EQ adjustments may be made as well as the addition of some dynamics-based signal processing, such as compression and limiting to bring the overall loudness level up.

As has been outlined throughout this chapter, mixing can be viewed as a complex balancing act. Although the mixing engineer has a lot of tools available when mixing a song or project, it is important to remember that a mix is limited by the tracks that were recorded, which in turn are limited by the composition itself. Therefore, a bad composition that was recorded poorly will never sound good in the final mix. A composition can sound truly amazing, as long as the engineer successfully balances each instrument in all spectrums.

DELIVERY NOTES FROM A MUSIC EDITOR/RE-RECORDING MIXER

David Bondelevitch, MPSE, CAS Having worked in the film and television industry as both a music editor and a re-recording mixer, I have suggestions for composers or scoring mixers who are delivering materials to the dub stage (also called mixing stage) for the final mix of the film (also known as the re-recording mix)

Figure 6.34 Mix Room

For composers trying to emulate an orchestra using virtual instruments, please learn the ranges of the real instruments. On a keyboard, I can make a tuba play C8. It would sound ridiculous. Writing outside the normal ranges will subconsciously inform your audience that the instruments are not real. If you want your clarinet to go lower, use a bass clarinet sample. If you want a bassoon higher, consider using an English horn

I will define a few terms as we talk through the process. First, a scoring mixer is the person who mixes the composer's music and delivers it to the music editor, who will in turn deliver it to the re-recording mixers on the dub stage after the director's approval. This mix may all be married into a single 5.1 file, or even a single stereo file, but typically the re-recording mixer will want more control during the final mix and requests what are called "splits" or "stems." (I don't like the term "stems," that has another meaning to mixers; they are what are created during the final mix.) Hans Zimmer delivers multiple 5.1 splits to the stage; strings, woodwinds, brass, percussion, and synth tracks (and sometimes more) are all recorded and delivered separately as 5.1 mixes.

Figure 6.35 Extended Range Button

Figure 6.36 Mixer Busses

This benefits the composer by making sure the cue makes it into the film in some way, rather than setting levels to the lowest common denominator, burying your music in the mix (or dropping the cue entirely)

Given the choice between one 5.1 mix of all the music, or multiple splits in stereo, I would rather have the control of splits. I can always turn it into surround using various tools on the dub stage. Many beginning composers are scared of delivering splits either because they think the music editor or mixers will destroy their intentions, or because they are overwhelmed technically trying to create splits. Setting up a studio for complex mixes (especially in 5.1) is sometimes out of people's budget, but since even local commercials now want 5.1 mixes, you may at some point need to upgrade your studio

When done mixing a cue and listening to playback of your mixes, be sure to listen with the dialogue up. Many of the changes made on the dub stage are to keep the music out of the way of dialogue. I would also make sure any soloists, particularly with instruments not typically used in an orchestral setting, (such as ethnic instruments), be on their own 5.1 split. It might seem silly to give a duduk 5.1 tracks, but any time an offbeat instrument is used, someone will question it.

Figure 6.37 Virtual Surround Studio in Waves' NX Processor

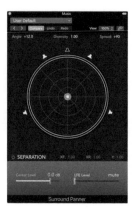

Figure 6.38 Center Channel

Putting it on its own track allows the re-recording mixer to equalize or add reverb to make it blend well with the other instruments.

I know some composers do not use a scoring mixer and attempt to mix their material themselves. I would not recommend this. Chances are your ears want to hear the music very differently than it will end up in the film. Experienced scoring mixers know not just the technical aspects of delivery, they will also know a how to create a good balance and proper splits.

Remember, for the splits to be useful, they must be usable on their own with nothing else. For the movie *Midnight in the Garden of Good and Evil*, Clint Eastwood decided on the dub stage to use *a cappella* vocals of k.d. lang singing *Skylark*. This would have been impossible if any of the other instruments were on her track. Perhaps the biggest technical issue is the use of reverb. Reverb on instruments must be on the same split as the instruments. You cannot use a single reverb on all instruments and give that on a track.

Reverb creates another problem. Many composers drown their music in reverb. Use of reverb in the scoring mix should be as little as possible, especially if you are delivering stereo files for a 5.1 mix. Reverb is one of the ways that stereo can be made into surround. Adding reverb in the final mix is also necessary to place the music further behind the screen, where it does not get in the way of dialogue. No one likes putting reverb on reverb

If you are mixing 5.1, please use the center channel sparingly. The center channel is where the dialogue is usually panned. Our ears use localization to interpret what we hear. If you put active solo instruments in the center channel, there is only one point source they will both come from, and that will make it more difficult for the audience to hear both the dialogue or soloist. In fact, some composers mix in "quad," not using the center channel at all. Many times, I have had 5.1 channel mixes come from the scoring mixer with an electric guitar panned to the center. To fix the mix, I use divergence to send the center channel equally to left and right, so that it will not interfere with the dialogue. (It's a bad idea to write a screaming electric guitar under voiceover.)

Figure 6.39 Divergence

In addition, use the LFE channel (low frequency effects, panned to a subwoofer) sparingly. The subwoofer can generally only play up to about 100 Hz, and there are not many instruments below that (other than bass and a few percussion instrument). I'm actually fine with the music delivered in 5.0 or 4.0, with no LFE at all. The re-recording mixer can always pan things to the center or LFE in the final mix if it is necessary

Time for another definition. The music editor is someone who assists the composer with the full process, from temp dubs to the scoring stage to the final mix. Music editors are often the only person on the music crew involved this long in the process. They may be on the show before the composer is even hired. As such, the music editor is often the translator between the director's vision and the technical aspects of music. If you are a composer and you are not using a music editor, you are really losing out on a lot of opportunities for your music editor to protect you. If you do not have room in your budget for a scoring mixer or a music editor, you (and your agent) need to be better at negotiating your package. (Most of the time, the music editor is separate from the music package; always ask up front if they plan on paying a music editor. You may get one without having to pay out of your pocket.)

A re-recording mixer creates the final mix of the film from the elements delivered by dialogue editors, sound effects editors, Foley editors, and scoring mixers to the dub stage. The re-recording mixers do this with adjusting levels, panning, equalization, compression, and reverb.

I would recommend that as composer, you take the music re-recording mixer out to lunch. More often than not, beginning composers are faceless to the final

Figure 6.40 LFE Level

mix crew. Make them your friends. Talk about the film, your score, and how you think things should play. It's an investment that will pay off later in the mix. Ask the mixer how they want the tracks delivered. Do what they request, even if you think it makes no sense to you. The mixers rule the dub stage.

I'm going to make another suggestion which will sound like the opposite of what I just wrote. The composer should never set foot on the dub stage until the film has been completely mixed and ready for playback. In 30 years of working on dub stages, I have never seen a composer make any useful suggestions during the mix. Most of the time, the composer's suggestions are basically "make the music louder." Remember, you do not own the music. You were commissioned to compose music by the studio or production company. The director gets to make all the creative decisions. Only the producers can overrule the director, but a good producer knows not to micromanage. Henry Mancini warned, "Leave your heart at the door before entering the dub stage." (I have seen directors ban the composer from coming to the dub stage. Don't be that guy.)

You will be represented on the dub stage by the music editor. Your music editor will call you if you are needed. Otherwise, all time spent by the composer on the stage will be very frustrating and may lead uncomfortable arguments. Chances are that when the final mix starts, the composer may still be recording or mixing tracks anyway. Your time is put to better use elsewhere.

After the mixing crew has mixed the entire film, there is usually a day for playback and fixes. You should attend this playback. Take notes (unobtrusively). After the playback, all the creatives talk about changes they would recommend. Let everyone else go first; they may address the problem anyway. If your comment is anything like "make this louder," you should choose your battles very carefully. If you keep saying that, no one is going to take you seriously. You can politely request a change, but you should understand that the director has every right to overrule you. In many cases, composers I have worked with leave immediately after the playback with no comments, or they give their notes to the music editor to discuss. Do not stay for fixes. Let the music editor and mixers handle that.

Immersive audio research

An important component in crafting a realistic orchestral reproduction is acoustic space imitation. This section explores the creation and usage of convolution reverb impulses as a primary means of mimicking real performance spaces. When using orchestral sample libraries, the source instruments are often recorded in concert halls with close and distant microphone placement which helps in the final mix by providing natural acoustic ambience. The techniques explored in this section are for situations where the source instruments are synthesized or recorded with little or no natural reverberation and yet the end production should sound like it is in a performance hall.

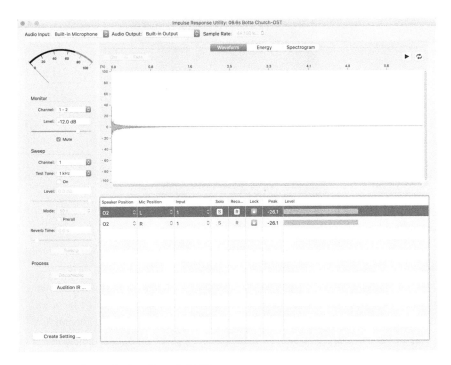

Figure 6.41 Convolution Reverb Utility

Panning

Understanding the process of panning individual sources lays an important conceptual foundation for setting up a convolution reverb to be successfully implemented. The first step is to understand the differences between stereo, surround sound, and 360° audio. Even though these three options are wildly different in format and capabilities, the basic principle of how to pan the sources remains relatively straightforward because of the goal of imitating an ensemble performs in an acoustic space. The reason mono is left off of the list above is because by itself mono is not panned or a panning destination, but rather a sound source which can be panned into one of the three options

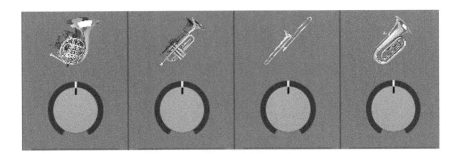

Figure 6.42 Stereo Pan

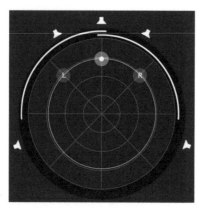

Figure 6.43 Surround Pan

Stereo

Panning in stereo involves two channels which remain in a fixed position. Creating a realistic stereo image means placing sources in realistic positions and it helps to imagine the instruments on a stage. While sources should not typically be panned hard left or right, instead place sources to match stage locations. Stereo images are more crowded than the subsequent multichannel options, which translates into the need for careful panning, adjusting levels to enhance depth, and carefully mixing the reverb to fit

Surround

Panning in surround is very similar to panning in stereo because the instruments are placed primarily in the L/R portion of the image, which represents the front. It is possible to expand further to the left and right extremes, but the surround channels are reserved for special situations and for reverb/ambience

360° audio

As with stereo and surround, sounds sources in 360° should still most often be placed in the front of the image, with exceptions only when special circumstances require it.

Figure 6.44 360-Degree Pan

While it may seem fun to place sources all around you, the primary objective of this chapter is to recreate natural and realistic sounding performances. 360° audio requires very specialized equipment for playback since sound would be coming at the listening position from all sides, and this equipment is described later in this chapter.

General panning guidelines

While there are no firm rules about where to pan instruments, there are some guiding principles to creating a successful image. The most important place to start is with an idea of your end goal for the image of your project. Do you want a realistic orchestral representation? Do you want something larger than life? Do you want the orchestra to surround you?

Stereo panning and depth

Even though there are only two channels to pan between in stereo, it is very helpful to think about panning having both width and depth. The side to side placement is accomplished with the pan control, and the depth comes from adjusting levels and effects. Some workstations have comprehensive panning mechanisms which allow you to pan and adjust levels simultaneously by dragging a single point around a grid. There have even been examples of effects which let you place multiple instruments in the image using multiple techniques to mimic the placement of instruments in an orchestra. As sounds are placed in the stereo field it requires more than simply adjusting position and levels to create a realistic image. If there is a trumpet player in an orchestra that plays from the back of the stage but is then moved to the front row, then the sound will change from the perspective of the audience. The trumpet will sound louder, brighter, and with a lower reverb to direct sound ratio. If these changes are mimicked in the workstation, then you can move instruments around the stereo image in a very realistic manner. Keep in mind that a sound is never just heard in one ear in an acoustic space, and so panning something fully into one channel of a stereo signal creates an unrealistic experience for listeners.

Surround panning

Placing sounds in a surround image is both easier and more complex than when panning for stereo. There are more options for placement angles, which can mean less processing is needed to make everything fit together, but the complexity of a surround system far surpasses a standard stereo system. A minimum of six channels with all associated equipment is required instead of two channels, a simple interface, two speakers, or possibly headphones.

Figure 6.45 Advanced Surround Features

There are two extremes when panning in surround. The first is to place all primary sources in the front three channels, as if they were on a stage with only ambient sound and reverb in the back. On the other end of the spectrum is placing sources in all directions to create a fully immersive mix. Of course, there are many variations possible in between those two extremes, with no rules on how surround panning is set.

The same principles described above concerning stereo panning also apply in surround, with placement, levels, and processing all contributing to the perceived distance and location from the listener. The listening position is one of the more difficult components when dealing with surround sound due to the equipment needs of monitoring in surround. Recording studios and movie theaters are the most common places where surround systems exist, although there are many home theater options. Acoustic design of the listening environment also plays a role and is often far more expensive than the equipment itself. Most home theater amplifiers can analyze their rooms and compensate for acoustic flaws, but all of these options require specialized installation and are generally expensive to do correctly.

360° panning

Panning in all directions is similar to surround panning in terms of concepts and implementation, but even more difficult to set up with speakers. It turns out that speakers directly above and below the listening position are complex to install and rooms which would be able to handle them are hard to design. Phantom images between speakers that simulate vertical positioning are a possible solution but not effective when designing for multiple listening positions.

Workstations

Most of the top tier workstations do not have 360° panning built-in and so third-party solutions are the only solutions. The problem is that using a plug-in to pan in 360° is inefficient and often problematic with DAWs such as Logic Pro X and the basic version of Pro Tools.

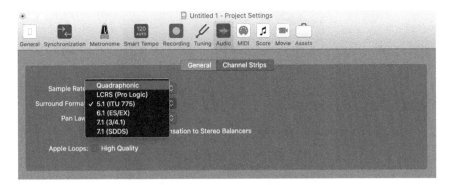

Figure 6.46 Setting Logic to Quadraphonic to Prep for 360°

360° techniques

Just as with surround sound, 360° panning has two extremes with front versus fully immersive panning and everything in between. The vertical axis adds additional options for panning and is a step up in terms of immersive audio, but there are no strict rules on panning instruments. Best practices suggest that the most important things should be panned to the front and that vertical panning is tricky due to poor vertical localization by human ears.

Monitoring options

Monitoring stereo, surround, and 360° audio all come with different challenges. There is a balance of working in the ideal format for a high-quality production and the ideal format for public consumption. 360° audio provides some of the most interesting options for realism and creativity but is currently one of the least utilized formats by consumers.

Waves NX

Instead of surround sound speakers or even a 360° speaker configuration, the biggest breakthrough is a simple device that attaches to the top of headphones and tracks head movement. The movement data is used to control a plugin which is able to reproduce sound directionality through the conversion to binaural audio and adaption through psychoacoustic principles. In other words, if the head turns in a complete circle then the image sounds like it is stationary around the head creating an immersive result. Sound can be panned anywhere in the 360° field, including up or down, and a pair of headphones can be used to monitor it.

Figure 6.47 Waves NX Head Tracker

YouTube and ambisonics

The same fundamental technology that is in the Waves NX device also exists in smart phones and can be used in the same way with 360° audio. The YouTube app is compatible with ambisonics and therefore capable of 360° audio/video playback. The app uses head related transfer functions to convert the audio from ambisonics to binaural stereo. Some web browsers can also be used to view 360° material, but the decoding uses a different process altogether.

The reason that an ambisonics capable app is important is because phones can be used with virtual reality (VR) headsets and provide a fully immersive experience to millions of people who already have such devices in their pockets. You can strap a phone to your face and as you rotate your head the audio image rotates as well. This, combined with tools like NX, means that 360° audio is both easily mixed and distributed using widely adopted tools

Recorded vs created

There are many combinations of techniques which can be used in a production. An instrument can be recorded with a single microphone, a stereo microphone pair, a

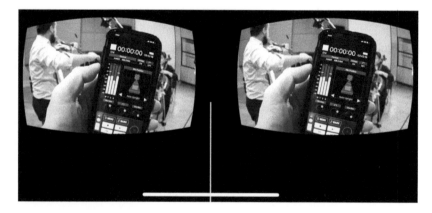

Figure 6.48 YouTube App

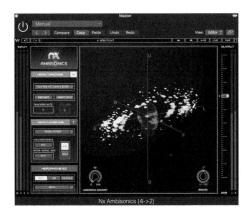

Figure 6.49 NX 360° View

surround microphone, or even an 360° microphone array. A project can have multiple source types all sent to a specific output channel configuration, which is both flexible and tricky; flexible because it means you can combine any types of audio, but tricky because creating a realistic end result when mix-matching materials takes some effort.

Start with the tracks that represent the big picture and fill in with the remaining pieces. For instance, if you record a full orchestra with a 360° microphone and then add other parts to it, the additional parts would be panned to fit within the overarching image.

360° convolution reverb

While reverb is covered previously in the section on effects and processors, it is worth exploring how 360° plays specifically into the creation of realism in orchestral productions. Convolution reverb combines the sonic analysis of one audio file onto another, menacing you can record the sound of an acoustic space and then put any sound effectively into that same space. There are several different ways to capture the initial impulse which contains information about the space, including making a loud full spectrum sound in the room while recording it or by playing a sweeping sine wave in the space and using software to create the impulse. The loud sound is often a starter pistol or popping a balloon, of which the latter seems to be most widely accepted as both easy, high quality, and less objectionable to venues than starter pistols. The sine wave method has a lot of benefits but is difficult in large spaces due to the need for large and well-powered speakers

Once you have an impulse you can load it as an audio file into a convolution reverb plug-in and route Instruments through it to make them sound like they are in the original space. Some plug-ins, such as Logic's Space Designer, are designed to work with ambisonics impulses. An impulse recorded in a room using a 360° microphone captures the sound of the room from all directions. When applied to an instrument in a mix and subsequently monitored in VR, the reverb surrounds the listener as if they are in the original performance space. Excellent virtual orchestral Instruments, such as the

Figure 6.50 Balloon Popping

Vienna MIR Pro, rely heavily on convolution reverb techniques to build a realistic sounding orchestra.

WORKFLOW

Should you want to create your own 360° impulse for use with an ambisonics capable convolution reverb, the following is a set of steps to follow and an accompanying video is posted on the companion website.

1. Identify the space for analysis.
2. Prepare necessary items:
    1. Balloons
    2. Stand to attach balloons to
    3. Pole to pop balloons
    4. Needle to attach to pole
    5. Tape
    6. 360° microphone with all necessary audio recording tech
    7. Audio editor
    8. Convolution reverb
3. Place inflated balloon at the desired source position. This could be at a central location on the stage or at typical locations of the Instruments of your choosing. For instance, you could place the balloon at every seat of an orchestral set-up and create impulses for every position.
4. Place the microphone at a good listening position, but it may be worth doing this multiple times to find the best position.
5. Pop the balloon, while standing apart from it. If you hold the balloon while popping it the some of the impact will be absorbed.
6. Edit the recording by trimming the beginning and the end.
7. Export the impulse and import it into a convolution reverb plug-in.
8. Use the reverb on 360° projects.

Figure 6.51 Prague Castle

Sam McGuire's 360° research in the Czech Republic

The soul of the music exists in the expression of the musicians and in the story that leads to the performance. One of the biggest limitations of a virtual orchestra is that one person is trying to compensate for the lack of community engagement that occurs when an ensemble plays together. It is a switch from many voices playing as one, to one voice trying to imitate many but still trying to be one. It is possible to create a successful performance but there are other factors which must be relied on. It is useful to understand the places where the instruments came from and who performed them. Just because a sampled instrument can be played one note at a time doesn't mean it can replace a virtuoso performance. You can place a reverb on to an orchestra, but what makes that specific reverb the right choice? This question is something which is explored in this section and has been the subject of years of research and field testing; to see what can help bring the digital orchestration to life

Background

As a technical author I rarely use the first-person voice, perhaps in an effort to present my findings as fact or to distance my own opinions from the task at hand. In this portion of this chapter I am letting that personal rule slide in order to explain the heart of digital orchestration as I have come to understand it. I began collecting impulse responses a few years ago after using impulses from other creators for many years. I have always enjoyed the idea of adding space around a sound from a specific place; someplace that actually exists and where music has known its halls. What started as a hobby to discover how the process works turned into a journey that will continue on long into the future. It can be overwhelming to create an impulse response, with so many variables and difficulties in planning, but I was fortunate to find a few real-world laboratories that offered me a chance to explore.

Figure 6.52 Charles Bridge

Czech Republic

I have had a connection to the Czech Republic for more than 20 years and the music heritage that grew in its cities and villages. When I first walked the streets of Prague in my early twenties, I knew who else had been there, but it didn't really sink in. I saw concerts in the same hall where Mozart had premiered Don Giovanni. I visited the hamlet where Beethoven's Immortal Beloved was rumored to be staying when he wrote to her of his passion. I had ridden a train past the birth home of Antonin Dvorak countless times and I hadn't even realized it, but I had seen it and wondered who lived there. I had seen the castle on a hill overlooking the Moldau river which inspired Wagner to compose one of his most famous operas. There are a thousand more examples of this country's place in orchestral music, and for many years it had been a part of my life without actually being a part of my life

Kutná Hora

In 2017 I decided to finally return to the Czech Republic after nearly 20 years away and I began to realize that my research into convolution reverb could be supported in a significant way in this amazing country. I started making serious plans and in August 2018 I returned and began the process of capturing stereo impulses in the St. Barbara

Figure 6.53 St Barbara

Figure 6.54 Karlštejn Castle

cathedral in a city called Kutná Hora. I arrived just as the sun began to set and was given access to the church after it closed to the public. Construction began over 600 years ago and it is an amazing church where people have worshipped for hundreds of years. This is where the seeds were planted about the true impact of a location and its history of being used. The impulses captured in this location mean something to me and to the music I use them on. It brings something to the performance and the emotional connection between the virtual instruments

Karlštejn Castle

In January of 2019 I was able to return and capture my first impulses using a 360° audio recorder. This next step was planned carefully because once I realized what was possible in a stereo format, I had to know what an immersive format would be able to accomplish and if it would bring additional realism to the final production. I was able to record impulses in the Martinů hall at the Academy of Performing Arts in Prague, which is a beautiful hall perfectly suited for orchestra, organ, and chamber music. This particular hall is housed in Lichtenstein Palace, which has a long history and I was fortunate to be given access for acoustic research. This set of impulses are used on nearly every personal project because it adds some intangible heart which I never want to replace. The trip also included something magical which was unexpected on a morning trip to Karlštejn Castle, the mountain home of Charles IV, Holy Roman Emperor and King of Bohemia. I arrived before the castle opened thinking there might be a chance to record some impulses but with zero connections or specific plans. With no other tourists on the first tour of the day, the guide agreed to let me pop a balloon in the largest room of the castle where concerts often take place and where the daily lives of castle residents would've existed for hundreds of years. I'll never forget the tour guide's eyes when she realized what I was about to do but she thankfully let it proceed.

Figure 6.55 St Bartholomew in Kolín

Study abroad

As sometimes happens with this type of research, things began to snowball in 2019-20. I brought twelve students on a study abroad in May of 2019 and we were given access to Střekov castle where Wagner had stayed and was musically inspired. We recorded impulses in the philharmonic hall in Hradec Kralove. We recorded a few in the village cathedral in Kolín, and also returned to Kutná Hora for a full set of 360° impulses in St Barbara and its ossuary. All of these were impactful experiences and the collection of meaningful spaces started adding up

Nelahozeves

The birthplace of Dvorak is someplace I've visited before – on a rainy day, years ago, I got off at the wrong stop and had to walk a few miles to the next village. I loved seeing the same woods that Dvorak would've known and walking along the same river he would've swum in. In January 2020 I was invited to visit his birth home and the church he would've been baptized in and heard music for the first time. The day I arrived I was also invited into the castle overlooking the village to record additional impulses in their

Figure 6.56 St Andrews in the Village of Nelahozeves

Figure 6.57 Nelahozeves Castle

Knights Hall. It just keeps going and going, with more people looking to see what the results are and how their spaces compare to others around them. The defining moment for me in the journey was meeting the descendant of the Lobkowicz family line and discussing his ancestor's support of Beethoven and the arts in Europe. We are currently discussing future projects to analyze the acoustics and create impulses from a whole slew of additional locations

Why?

The point of this narrative is to tell you about my personal journey to find the heart of my own digital orchestration. I've talked to a lot of people who use instrument libraries, and each is more passionate than the next about what works for them and how amazing it is. Find the angle about which you are passionate and dig in as deep as possible

Immersive audio

In spite of the trips to the Czech Republic and capturing impulses in castles and concert halls, there were a few other important discoveries. In the last trip there was another objective which was to record musicians in stereo and 360° and do a full analysis of how they sound and more importantly how the immersive audio affects the listener's perception of realism. There were two key experiments with separate goals. The first was

Figure 6.58 Knights Hall

recording individual members of the Prague Castle Guard/National Police Band in their rehearsal facility, which were then replicated using sample libraries. Samples were played with and without reverb captured in the same rehearsal space. The second experiment was to record the Wihan String Quartet in a recording studio using both stereo and 360° microphone configurations, with the goal of determining which configuration is generally more pleasing to the listener. It turns out the primary piece recorded was a Beethoven quartet, which has now been placed virtually in the Knights Hall at Nelahozeves Castle due to the impulses captured there. Preliminary findings suggest that immersive audio helps with the realism and musical enjoyment of orchestral libraries, but sadly COVID-19 interrupted the testing process. Additional research will continue on and once the pandemic is under control then listening tests will progress to determine increasingly definitive results.

Chapter 7

Orchestration stories and workflows

In this chapter, musicians and composers weigh in on specialized topics relating to the digital orchestration process. Each story was hand-picked in order to add to the narrative of this text. Not every point of view supports digital orchestration as the best option, and certainly some versions ended with a real orchestra recording the parts.

Sue Aston: digital orchestration

Composer and violinist Sue Aston has appeared on classical recordings, radio and television, both nationally and internationally, and worked with eminent musicians such as Simon Rattle, Nigel Kennedy, Peter Donohoe, Yehudi Menuhin, Sir Charles Groves, Esa Pekka Salonen, Gordon Giltrap and Chris De Burgh. She has also supported the folk legends Martin Carthy and Dave Swarbrick. The BBC1 *Heaven and Earth Show* presenter Simon Calder interviewed Sue about her composition work on one of their episodes about the Cornish landscape. Sue's solo videos have been broadcast on the Sky TV music channels Classic FM TV and OMusic.

Discography: Sue has released three solo albums *Sacred Landscapes*, *Inspirational Journey*, and *Between Worlds*, and a recent EP which features solo piano with strings and guitar – *Winter Keys* – and the DVD *Reflections of Cornwall*.

I use the Cornish landscape and nature to help compose my music. The stillness and background sounds of Cornwall weave through my music. Music can be the soundtrack of an ordinary day and an extraordinary day. The movies that play in our minds can conjure up their own soundtracks. Music can change the atmosphere of a room or a place or a frame of mind. Although I use folklore and legends and aspects of my own life to trigger each composition the instrumental aspect of my work is written so that the listener can weave their own images and experiences into each piece.

Sue Aston has also featured on the BBC Radio 4 documentary, *Derek Tangye: The Cornish Gardener* where she was interviewed by John McCarthy about her work. She has also been interviewed by BBC Radio 3 about her interest in English composers who were also influenced by the landscape. Radio 2 featured Sue's track *Initial Bond* for a program about composing music for a loved one who has died.

With advances in technology regarding the sampling of orchestral instruments, many composers use the sound of synthesized strings within the backdrop of a virtual orchestra when writing music. This obviously saves time and money. If the music score is busy, then the synthetic string sound can be very convincing, but in areas where the strings are exposed, especially during a solo section, then it is worth investing in hiring real string players to play over the track, to add an authentic performance.

In particular, certain aspects of string performance and technique reveal whether or not a piece of music features real string players. Different registers of a string instrument can sound thin, especially at the extreme high and low end of the range. In combination with the use of vibrato, which is often too narrow and fast, this can really highlight the fact that the sound is produced electronically. During the performance of a single note, a string player will create a multitude of different sounds by changing the bow speed and varying the vibrato. The sound of the note will grow, by building up to a crescendo, then fading away with a diminuendo.

Synthesized strings don't offer the scope needed for sensitive playing with the bow, and techniques such as *flautando*, *sul tasto*, *sul ponticello* and creating natural and artificial harmonics can all sound contrived. Pizzicato can sound too harsh, snap pizzicato and *col legno* can sound far too percussive. Tremolo can especially sound too rapid and unconvincing.

The very aspects which make a string instrument sound beautiful are missing – in a real performance the notes are created by the player carefully balancing the bow with an even weight between the fingers, then bowing the hair over the strings, which in turn vibrate and manifest the sound out of the depths of the varnished wood.

From my own perspective, when I have been asked to play in a recording studio over an electronically produced track, factors to take into careful consideration when performing are that a cold wooden instrument may take time to warm up, and special attention is required regarding intonation.

String players traditionally tune by ear and listen to the interval of a perfect fifth, whereas a synthetic string sound will be too bright and artificial sounding. Vibrato needs to be used mindfully, as if it is too wide, this can alter the pitch slightly and clash with the perfect tuning of the electronically produced music. Playing to a click track doesn't allow for rubato, and so it is difficult for the music to naturally breathe and flow.

Despite these differences and challenges, using real string players enhances any music score, by adding a depth and rich quality of sound. You may assume that you would need to employ a whole string section to achieve this result, but I am often hired to play and record multiple layers of violin and viola parts.

Real strings add an authenticity to what can sometimes be a one-dimensional, flat sounding layer of electronic music. It creates a perfect balance between the use of traditional string instruments and modern musical recording techniques.

Col. Vaclav Blahunek: wind symphony orchestration

> Col. Blahunek is the Director and Chief Conductor of the Prague Castle Guard and Czech Police Symphonic Band.

1. What should a composer know before they compose music for a wind orchestra?

I would expect these absolutely necessary skills of composers for wind orchestra. They should know:

- Tone ranges and sound possibilities of every instrument.
- Dynamic scales of every instrumental section.
- Technical difficulties of wind instruments, like breathing, slurring, articulation, attacks etc.
- Forms of music.

Figure 7.1 Col. Vaclav Blahunek, Ph.D.

- Rules of counterpoint.
- Concepts of musical styles.
- To know "how to score music for winds" = experiences and research into wind band scores.

Special features:

- Unique compositional techniques.
- Exploring new sound qualities and expression of winds.
- Easily recognizable personal style within almost every measure of music.
- Magic of musical speech: "to know what to say and how to say" in which case understanding of listener makes no difference for it!

2. What mistakes do composers make when writing for a wind ensemble?

The most frequent mistakes of wind band composers:

- Lack of respect in the balance of sound (it is usually the main responsibility and creativity of a conductor).
- The blend doesn't work.
- Too many harmonic mistakes (sometimes typos due to the publisher).
- Missing respect of special wind characteristics of overtones.
- Pieces of music are too long – they do not know where and how to end the piece.

3. What mistakes do arrangers make when transcribing music for a wind ensemble?

Mistakes of arrangers of symphonic scores for wind ensemble include:

- They do not respect tone ranges.
- Patterns of process, e.g. strings to clarinets, cellos to saxophones.
- Does not respect harmonic progress of composition, they often change notes!

The best way to orchestrate music is to "decompose" the piece, find the main structure, uncover the roots and then match it to totally different sound colors, suited to the wind medium.

Don Bowyer: composition/orchestration process

Prof. Don Bowyer is Dean of the School of Arts at Sunway University in Malaysia. He has played trombone in more than 50 countries and has published more than 60 compositions that have been performed around the world.

Composition biography/use of technology

As a composer, almost all of my work has been for a specific group of live performers. As such, my primary technology usage has been notation software and a MIDI controller, focused on printed notation much more than audio output. I should also mention that I am an active performer on trombone, which I suspect has had an influence on my compositional style and process.

I started composing in the late 1970s, using pencil and paper, later copied in ink. I do not miss those days, although there was a period in the 1980s when I had a decent "side gig" as a copyist. When Finale 1.0 came out in 1988, I got very excited and decided I needed to switch to computer-based composition. While I was (and still am) a Windows user, the first version of Finale was Mac-only. Fortunately, I had a Mac SE in my office. By the time I left that job two years later, I had switched to the Windows version of Finale.

I still remember being intimidated by the unboxing of that first version of Finale: three spiral-bound instruction manuals, each an inch or more thick. I was just OCD enough to actually read them all the way through, but it still took years to feel proficient. My other clear memory from those first days was w-a-i-t-i-n-g for the screen to redraw. It literally took several minutes, sometimes even ten or more, to redraw the screen for a full score on that Mac SE. In the early days, we learned to edit without redrawing the screen, then redraw after several edits. Of course, that meant remembering what changes you made in one part as you edited another.

I did not have a MIDI controller in the early years. Even so, I quickly discovered that Finale's Speedy Note Entry was a much faster method of note input for me than the Simple Note Entry. I now use a MIDI controller for almost all input, but still with Speedy Note Entry. I have had several miniature controllers I liked, recently including the 25-key XKey and the Bluetooth-enabled NanoKey Studio. My computers have all been laptops for at least ten years, so the smaller keyboards enhance that portability.

Because my ultimate goal is typically printed music, I primarily use playback only to identify wrong notes, as in "I don't think that was what I wanted." As such, I have mostly used the built-in sounds. Recently, however, I upgraded to Note Performer because the Covid-19 lockdown forced me to perform a few times with Finale files. I like the string and piano sounds a lot more now. On the other hand, after performing through Facebook Live, delivered over phone speakers, I'm not sure anyone else noticed!

My earliest compositions were almost entirely for jazz groups – both big bands and combos with two or more horns. In more recent years, nearly 50% of my composing is for "classical" soloists or ensembles, both large and chamber.

Orchestration process

Besides composing, I have also done a fair amount of arranging or orchestration (which I see as the same thing, though I know that some do not). For both jazz and classical work, my compositions usually start with melodies or, occasionally, a rhythmic motive. The harmony almost always comes later in my process. In other words, I usually compose horizontally, then orchestrate vertically. As such, the orchestration work for me is pretty much the same whether I am composing my own work or arranging an existing work. To be clear, I am not advocating for this approach – I suspect most of my favorite composers have a different process that blends the harmony and the melody into a more symbiotic whole. For me, however, single-note melodies come more easily. Perhaps this is influenced by my years as a trombone player (and my lack of keyboard skills).

Jazz orchestration

Most of my jazz compositions are for big band. As a graduate student I had an arranging class with a fellow graduate student, Dan Gailey, now the long-time jazz director at the University of Kansas. Among other things, he shared with me a voicing chart, based on Basie voicings, that he said he got from Tom Kubis. The chart included suggested four- and five-part voicings, both open and closed, for every scale tone in major, minor, dominant, diminished, and altered chords. In my early days, I spent a lot of time with this chart, meticulously duplicating the Basie sound. Of course, I eventually internalized this enough to "break the rules" and develop my own style. I no longer know where that chart is, but I am sure that I still use the same sounds more often than not. Applying this, if I have a melody in the lead alto, the other four saxophones flow down from that. Then I go back and look at voice-leading, particularly repeated notes in an inner part, and make adjustments to smooth the lines. The same process holds for each horn section. When combining sections, I use a couple of different approaches. Trombones, for example, might duplicate the trumpets an octave lower. More often, though, in a section with all brass, with or without saxes, I will try to create an interesting bass trombone line instead, filling in the horns between using the same four- or five-part principles, but spread over eight or ten brass.

I should mention that I also use a fair amount of unison lines in horn sections, sometimes with all horns, and sometimes only within the section. One sound I possibly use too much is writing simultaneous independent unison lines for each section.

My approach to the rhythm section probably grows out of my background as a horn player. For almost every band I write for, I believe the bass player will come up with a better bass line than I will write. The only exception to this would be a specific repetitive bass line that I might want. Likewise, I believe the keyboard and guitar players will provide better voicings than I will, so I usually write chord charts for them. Incidentally, I rarely write rhythm parts so I can hear them in playback.

For drum parts, I will indicate style and tempo, and when to play time. If there are specific hits I want the drummer to play, I will write those in the staff (the rhythms, not the specific instruments). The last thing I do with the drum part is to write in horn cues for most of the chart. I put this in a separate layer in Finale, above the staff. I create the cues by copying the rhythm of the lead horn at a given moment. When writing for smaller jazz groups, I use the same principles as the big band, but with fewer horns.

In summary, my approach to jazz composition and orchestration is performer-informed. Jazz has been primarily a performer-based art form. I consider the relationship between composer and performer to be a collaboration.

Classical orchestration

Whether because of my jazz origins or my performance career, I have the same collaborative approach to classical composition and orchestration. I don't write improvised percussion parts (unless it is an aleatoric piece), but I do welcome performer interpretations that don't match my original conception.

My compositional approach in classical music tends to vary from my approach in jazz, particularly in chamber music, where each instrument will be more independent and, therefore, less "orchestrated." My large ensemble pieces, though, have a similar orchestration approach to my jazz process. The harmonies and voicings are usually different, but the sectional approach is similar.

Additional uses of technology in my compositions

As an aside, I have become very interested in recent years in aleatoric music that involves the audience in some way. This is not really an orchestration process, but it does bring up a couple of other examples of use of technology.

One piece that I have performed fairly often is for trombone, multimedia, and audience cell phones. I wanted to involve the audience in some meaningful way in a piece that serves as a metaphor for a day in my life. I composed individual sections of music for trombone and electronic sounds. The electronic sounds are in a Flash timeline (not a recent composition) that includes animations projected on a screen. The individual sections are titled for activities in my typical day: administration, composition, performance, technology, teaching, and family. A seventh section, called unplanned demands, interrupts the music every time my cellphone rings. Following a short interruption, the music picks up where it left off. My cell number is always visible on the projector screen and the audience is encouraged to call as often as they "need to."

In other technology uses, I have composed a couple of pieces for instruments involving a looping app on my phone. Finally, I have composed a couple of small film projects, synchronized through Ableton.

Tip for writer's block

In closing, I thought I would provide a tip for overcoming writer's block. When I am struggling to get started, I tell myself to write one note. Then write one more. Repeat until something sounds good, then go back and delete (or edit) the parts that don't sound good. This process is so much easier with notation software than it ever was on paper!

Rahul Shah: library analysis

Rahul Shah is a Canadian composer for film, television and multimedia from Toronto, Canada. His sound has often been described as classical/orchestral crossover with world elements. Rahul has written music for projects distributed on Netflix, Amazon, PBS, BBC, ITV, Discovery and worked on advertisements. He has gained professional experience in composition, orchestration, editing, score preparation, sound design, and arranging. In addition to his own freelance projects, He has assisted composers for film and television at Hans Zimmer's Remote Control Productions, including Henry Jackman (*X-Men*, *Captain America*, *The Interview*, *Jack Reacher*, *Kingsman*).

There is no shortage of sample library developers on the market, just naming a few we have VSL, Spitfire, East West, Orchestral Tools, Cinematic Studio, and Cinesamples. All of these libraries and sample developers pose advantages and disadvantages to their software, but sample libraries have come extremely far in the last 15 years. Much like cell phones just the fact that a smartphone exists is conceptually amazing. I've heard stories of composers from 20–30 years ago running sample libraries off of floppy disks and other storage mediums far more primitive in comparison to the hard drives we use today.

Let me be clear, I appreciate many different libraries from all of these sample developers. They do great work and deserve for you to buy their libraries. In saying that, I'll share some of the comments I've heard in the industry about certain libraries. People have argued in the past that Spitfire, while it sounds great out of the box isn't always the best representation due to limited mic positions, articulations and a lack in dynamics. Some have argued that Spitfire is a very good, out of the box, solution but prior to BBC Orchestra it didn't provide as realistic an implementation of an orchestra. VSL (Synchron) has been mentioned, much like East West's PLAY engine, to be somewhat heavy on the CPU and RAM resources of your machine. Orchestral Tools features stud libraries such as Berlin Woodwinds and that is definitely my favorite choice for woodwinds (they also have dynamics such as *m*, *ff*, *f*, *p*, etc.) but there is no further control on quantity. Cinesamples does a great job recording, editing and managing samples of their libraries. However, if you're not going for a large sound they can be unnecessary for certain arrangements. With all of these libraries there is a reason why people still pay thousands and sometimes millions of dollars to record with the top orchestras in the world! You always need certain human elements (natural human imperfections in playing, breathing, air, the room), which create the true feel of an ensemble.

I'm sure everyone uses real instruments, however, for demo/mock-up purposes getting an accurate representation of your music is important (e.g., dynamics, articulations and feeling as close to a live performance). There are situations where a live ensemble/player isn't available or an option – for this reason, I really like the libraries from VSL (Vienna Symphonic Libraries) if that is your main goal.

Their libraries can also serve you well in a piece written for concert. For the most part they have the dynamics and articulations you'd need and the samples are recorded well. Top educational institutions in composition or film scoring recommend or provide VSL as a bundle to learn command of an orchestra and hear realistic implementation of what you've written. VSL does a great job of nailing the exact orchestration and timbres (e.g., three horns sound like three horns) rather than only having access to a solo horn, or two horns or 6/12-horn patch. VSL has done a great job of making sure that they are placed well in the room so that if managed properly, instruments or sections can provide the illusion of true orchestral panning/ensemble. To translate this, it sounds like your music is being playing by humans and after you sequence the patch in your DAW (e.g., Cubase, Logic, DP, Pro Tools) they're realistic and sound like a human could've played them.

For a composer who wants to have the most realistic simulation of what they're writing so they can hear what their ensemble will sound like, Vienna Instruments/Synchron-ized series (Volumes 1, 2, 3, 4) provide the most accurate picture to them. When orchestrating it's extremely important to know what the weight of your ensemble is, you don't want to use a 6 horn or a 12-horn patch when your ensemble has two horns because you'll wonder why your horn sound isn't as strong. When you combine the exact dynamic references that are available in Vienna (e.g., *mp*, *f*, *p*, *ff*, *fff*) this going a step further to make sure you are hearing exactly what you or someone else has notated in a score. Similarly, if you have two clarinets and two flutes but you're using a patch with five to ten woodwinds each your expectations will not be met during a live recording. Additional parameters can be programmed in Vienna Synchron player such as Xfade (Velocity Crossfade or other MIDI CC (control change) parameters. A big problem with orchestrating for strings has to do with numbers, these string patches in some libraries can have 10–12 violins in one violin patch. If you add a whole family by section you could end up expecting the sound of 50+ string players and in the actual recording you may only have 15. In terms of hybrid orchestral, because the sounds are so realistic when merged with live players from an original recording it forms a hybrid sound. Then if you need additional thickening of the orchestration for a purpose then one could add a bit of extra weight from the other libraries to thicken once the recording has been completed. This will need to be evaluated according to the context, an intimate chamber sound isn't likely to require as much thickening as an orchestral epic trailer or action scene. Not to mention there are countless instruments that are used in concert writing that aren't even sampled with the other sample developers just based on the fact that their prospective buyers wouldn't generally have use for these instruments in cinematic writing. Having these additional tools either for scoring or writing is definitely an added bonus over other sample developers and libraries.

Here are a few examples from the concert world and film scoring world where I exemplify how Vienna translates well from score, demo/mockup, live recording and then to the final dub stage. You can take the score from any films or concert, program it into your DAW of choice using MIDI and you're likely to notice Vienna Instruments/ Synchronized libraries provide the most realistic implementation of what you see on the score. The best-case scenario is getting a realistic sound with true illusion of ensemble as well as an epic sound that fits the context after thickening with other libraries to create a sample hybrid and hybrid orchestration along with a live recording. At Berklee College of Music one of the assignments I remember in the scoring program is sequencing these pieces/classic works to create mockups and create the illusion of an ensemble using MIDI.

pp

Figure 7.2 Excerpt from Pavane pour une infante défunte by orchestrator and composer Maurice Ravel

For instance, this excerpt by Maurice Ravel requires sounding like two horns – Vienna libraries do that and also allow you to select the *pp* dynamic in the patch. It is also important to not quantize the melody 100 percent according to a grid because then it doesn't feel as musical as the original piece. Allow for human pacing so that the integrity and phrasing of the piece is maintained. You can then rely on MIDI CC 1 (Modulation) to add some more expressiveness to the phrase and further shape the line. Samples are recorded for each dynamic layer (*pp, mf, f, ff, fff*, etc.) instead of being duplicated. It's important not to just use a two-horn patch but to use a solo patch as a separate performance to create the illusion of a small horn ensemble

This excerpt by Copland is a great example of how Vienna libraries can be used effectively for an orchestral piece while maintaining the sound of the brass section (four horns, three trumpets, three trombones, and tuba). You many need to layer using more powerful libraries for the desires sound but if your goal is to create exactly what is written on the page in terms of number of instruments, dynamic layers, realistic attacks/releases, human dynamics and articulations (e.g., *marcato, staccato*, etc.) then I think Vienna is a great place to start. It is also a great learning tool to hear how various numbers of an orchestra sound so that you aren't as surprised during the recording session when it comes to the perceived weight of your sound. This can also result in you orchestrating differently and approaching registers of the respective instruments in a different manner.

Ondrej Urban

Ondrej is head of the HAMU Sound Studio and the Department of Sound Design at HAMU

My preferred way of orchestral recording is to understand all of the components of the process: score, orchestra, hall. The most common practice is to build the basic re-

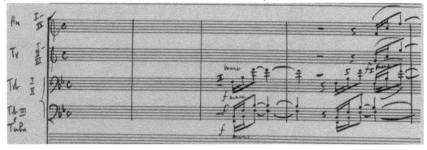

Figure 7.3 Handwritten excerpt from Fanfare for the Common Man by composer Aaron Copland https://loc.getarchive.net/media/fanfare-for-the-common-man-1

cording system around few mics at the beginning. It really depends on wishes and needs of the final recording - I mean, I can choose different setup for soundtrack movie and another one for making just sound recording for archive purposes, for radio or TV.

Starting with the basic stereo-mics setup really helps to understand what the sound of the hall is, projected to a basic system. I usually use ORTF, XY, AB or Decca 3 setup or combination of them, depending on the size of stage and orchestra. If the score is written well and the orchestra is playing well too, half of the job is done. Then, after listening to the basic stereo setup, I ask, what is missing. Often, I have musical director by my side, or I have the score for myself and I need to "read" it before recording. The missing components are usually the direct bass signals, presence of solo instruments, attacks of drums and percussions. Making the basses less ambient helps to build the sound sitting firmly on the bottom. Spot mic for solo instruments or orchestral solos (e.g., flute) helps for putting the solo voice in front of the filed, more "in your face." And spot mics for drums and percussions (mainly timpani and mallets) improve the attacks and perfect rhythm pulses. Another spot mic helps the harp to be more present, also horns will like it. So, after some adjusting, we have basic stereo setup plus a few spots, circa 12–20 tracks. My idea is to make the basic sound as good as possible only using stereo system and then to help with some spots. This approach works well with more classical scores (Beethoven, Dvořák, Brahms...) and with sound for radio and TV.

Another way of building the sound field is recording for movies, pop-oriented orchestral recordings etc. I go usually in the opposite way: I start with separate and more directional mics for groups of instruments – 1st violins, 2st violins, violas, cellos, basses – than woodwinds (two to four mics), brass and horns (two to four mics), percussion (two to five mics), soloists (one to three mics). Sometimes we have organ, choir and for example orchestral piano, celesta ... (so, a few mics more). I prefer more contact miking in such situations, so as to have more of the presence. But for "gluing" everything together, I use a combination of ambient mics (mainly AB) and artificial reverberation. The final sound is more "in-your-face," you are not limited by the dynamic levels of different orchestral instruments and their groups etc. It is this way that I usually do the soundtracks for movies or computer games. Depending on the needs and size of the orchestra, we record usually in the studio (drier) or concert hall (wetter). My favorite concert halls in Prague, where I do the orchestral recordings are: Dvořák's Hall of Rudolfinum, Smetana's Hall of Municipal House, Martinů's Hall of Lichtenstein Palace (Academy of Performing Arts).

Figure 7.4 The Rudolfinum Concert Hall

In movie soundtrack recordings, pop-music recordings and recordings for games we usually need to sync with the picture and/or pre-recorded material (sampled instruments, rock band recording, click etc.) In such a case, every player needs to wear headphones. The main difficulty is the cross-talk from players' headphones to each section mic. I usually adjust the level of click and playback tracks to be as low as possible, giving the players the right tempo, but not having too much bleed.

Camille De Carvalho

I only use real instruments when recording music. The reason goes back to when I was 16 – at that time I was just a young pianist and I had just bought an electric keyboard, wanting to connect it to my PC to play new sounds. I installed notation software and began writing with all the sounds available, which I had mostly never heard before. My family is not really into music so apart from piano, guitar, drums and maybe saxophone I didn't really know any other instruments. I also installed a lot of keyboard emulations: Minimoog, Hammond organ, Fender Rhodes … and I was quickly overwhelmed by the quantity of different sounds I could play. How could I ever choose between them?

My pieces were not progressing at all as I was constantly tweaking the knobs of my virtual Minimoog and rerecording everything. And then I met Stephane. He saw me playing piano in high school and introduced himself as a flutist. At first, I laughed at him: what a useless instrument! Monophonic? Only three octaves range? Why not playing piano?

"Not the same sound," he said.

I answered that I could do every sound I wanted with my computer.

"Oh yeah? Well then, let's try." He sent me a short recording not even using strange techniques of him playing flute and asked me to replicate it. I tried, I struggled with vibrato, with dynamics, with the erratic tempo, with the breath sound – I couldn't.

Then he lent me one of his old flutes and told me to try it, explaining the basics to me. I discovered that there is more to music than pressing keys. Finally, he invited me to a rehearsal of his orchestra. That's when I understood that the infinity of sounds you can make with a computer does not include the infinity of sounds you can make with real instruments. So I had to choose a side: I could stay with my computer and use these sounds, but I already knew I had trouble picking one sound, or I could try with real instruments where you have a limited number of instruments, a limited number of techniques, to see if it was better for me.

So, I began writing orchestral pieces but, as I had no orchestral culture, my pieces were unusual to say the least. Not a lot of place for strings, a lot of rare winds and a lot of parts were actually unplayable. I also found out that if you're not already known, no orchestra will even bother looking at your pieces and if you can't pay session musicians, you'll have a lot of trouble finding enough of them to record an orchestral piece. I decided to do it all by myself. Luckily, my parents moved to China when I was 18, so the first time I visited them, I headed to Music Street where you have musical instrument shops for miles and bought my first woodwind instrument, a clarinet.

Since then, I have spent a lot of money on instruments and lessons. I bought all the members of the woodwind family I could find, then went on with the brass family, then the strings. I took lessons for some of them and used the fact that there are a lot of similarities to learn the others. I think the best bet is to take lessons for the hardest instruments to play: oboe, horn and violin. Then you can play most of the instruments in the orchestra. Of course, I'm not an expert in each of these instruments and sometimes, I need to write some difficult parts I can't play. But ten years in different orchestras helped me meet a lot of people and it's really different to ask a violinist to play 16 parts of long notes during 15 minutes than to ask him to only play the nice solo part. But the most important thing is that while learning how to play these instruments, I learned what is idiomatic and what isn't. Nowadays, it's true that some libraries can emulate string instruments really realistically and you can approach most sounds (although still today, I never found any piano VST with a mute. I love the mute on upright pianos!). But something that is still noticeable is when you hear something an instrument isn't supposed to be able to play. Even if you're not a musician, it sounds wrong. I once heard a chromatic harp glissando for example: the sound almost convinced me, but I knew it wasn't possible (although some chromatic harps exist, but they're really rare). And learning all of this taught me how to write so it sounds legit. Also, I can't change my mind at the last moment and replace the oboe sound by a clarinet sound: every note I write must have a meaning, a reason, or I'll spend hours recording for nothing. Every time I listen to one of my pieces, I can say proudly say "I recorded these flutes one by one, little by little," even though even a mellotron sound would have done the trick. The satisfaction I get out of it is immeasurable.

Cory Cullinan, a.k.a Doctor Noize

Digital orchestration is not always about making the final project digital. There are times when I use digital instruments to actually render and perform the final recording of my arrangements for both Doctor Noize and other projects. But there are also still many times when you go with "the real thing" in the final recording. It's important to note, however, that almost nobody goes from conception to recording a large acoustic ensemble without digital technology anymore – in either the arranging or recording phase. Hats off to those brilliant purists who do, but honestly, even a genius like Mozart would probably use digital technology in almost every phase today. Most music copyists now work via digital means instead of by hand, too

There are many reasons for this. One of the main ones is that advanced scoring software allows you to extract parts from a full score instead of writing them all out again for each instrument. If you've never scored for orchestra yourself but have seen the film *Amadeus* you know how helpful this can be. In the film, Mozart is struggling against health and timelines to finish his *Requiem*. It's not the composing that's slowing him down; it's the time it takes to get his brilliant creation onto playable scores and parts for the musicians that's keeping him up at night and killing him.

Anyone who's ever written for a large ensemble – orchestral, choral, musical theater – can relate to this, even if it takes us longer to compose than it took Mozart. Mozart could write complex music straight from his head to the page (it was like he was taking dictation from God, said his rival in the film), but no amount of genius could reduce the time it took to then copy each instrument's part from the full score onto a new page. That was just grunt work.

Figure 7.5 Doctor Noize

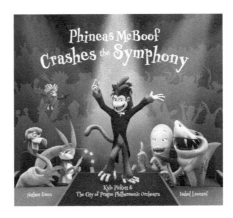

Figure 7.6 Album Cover

Advanced scoring programs like Sibelius and Finale take much of the grunting out of the grunt work. More than ever, with improvements over the last decade from software developers, full scores can be made into parts extremely quickly in a process called extraction. Most of us compose in full score mode, and then extract those parts into individual scores for violin, bassoon, bagpipes with distortion pedal, and any other parts you've written. (Okay, fine, I've never scored a piece for bagpipes with distortion pedal – but that's my loss and the world's, and frankly a gap in my production output that begs to be rectified.)

In my experience, it's a marketing myth that professional-quality parts can be extracted with the touch of a button in these programs. Almost always, several things that were laid out well in the full score don't transfer over perfectly – a dynamic too close to a note, a text box that now sits in the wrong place, or even a problematic page-turn for the player. But instead of taking dozens of minutes or hours to copy the part by hand from the score, you can generally make the proper adjustments to an extracted part in just a few minutes

A glance at this page from my Doctor Noize double album *Phineas McBoof Crashes the Symphony* shows just how much time and life expectancy can be salvaged from these gains. Mozart would be jealous of the resources at my and your disposal.

Figure 7.7 In the Studio

Figure 7.8 Live Performance

This project, which I recorded with the City of Prague Philharmonic Orchestra and 20 vocal parts (16 solo roles and the four-section Stanford Chamber Chorale) was a doozy to produce. Do the math in your head: The work is two hours and fifteen minutes long, with 20 vocal parts and over 30 instrumental parts. The full score is hundreds of giant pages – and that's *before* I had to extract all the parts – and contains dozens and dozens of musical numbers and scenes. Just writing that, or looking at my beautiful completed score now, makes the modern mind think: "No way I'm ever doing that by hand."

Even with this technology, it was a gamble for me on every level – financially, commercially, artistically. After a well-received tour of Doctor Noize orchestral shows introducing kids to the orchestra – started by an inspired commission I received from the McConnell Foundation to write a live orchestral show for kids, families and schools – we made the commercially questionable decision to make it into a recorded work so kids, families and educators who couldn't attend a show could still experience it. And we decided to make it in two full narrative acts, to give kids the experience of a legit opera, because we thought modern kids and even modern adults were smart. (Every night as I looked in the mirror working on this project, I realized that second assumption was a big leap …)

We had secured just enough funding to produce it through a digital fundraising campaign on Kickstarter that was ultimately profiled in detail in the European Business Review – we had a major orchestra and Grammy-winning opera stars Isabel Leonard and Nathan Gunn had committed along with our normal Doctor Noize cast. But two things that haven't changed since Mozart's day are the unknowns of whether the production will go off without a hitch and whether or not the public will like it.

Figure 7.9 The Dream Team

When you imagine that composers like Mozart and Wagner composed operas even longer than *Phineas McBoof Crashes the Symphony* with no digital technology whatsoever, and with their careers, reputations and livelihoods on the line, you suddenly gain an even deeper appreciation for the work of these masters. The production took me over a year of fairly constant work to compose and produce. But unlike the aforementioned masters, I had digital orchestrating software that allowed me to audition the ideas that came out of my head to hear if they sucked and change them before anyone else heard them; use quickly-exported digital representations of the score along the way in vocal rehearsals instead of piano reductions; demonstrate at any time to a potential performer, funder or future commissioner just how much work I had done, in pristine and beautiful score form; and, equally important for a composer's state of mind, easily create digital copies in multiple locations as backups. Imagine being two thirds of the way through writing Beethoven's Ninth or Mahler's Eighth after years of work, and going to bed every night knowing that the only copy that existed in the whole world could be burned down in a building fire at any time! It's enough stress to make a composer like Mozart die before he hit 40. Wait a minute …

So. Even with all the best digital technology in the world at my disposal, I *still* almost screwed it up. I will be the first to admit that digital scoring software is what bailed me out. It literally was the pivotal difference between a disastrously incomplete overly-ambitious failure and a glowing success story with many accolades that somebody just invited me to write a book chapter about. Here's how.

Figure 7.10 Trying to Stay Awake

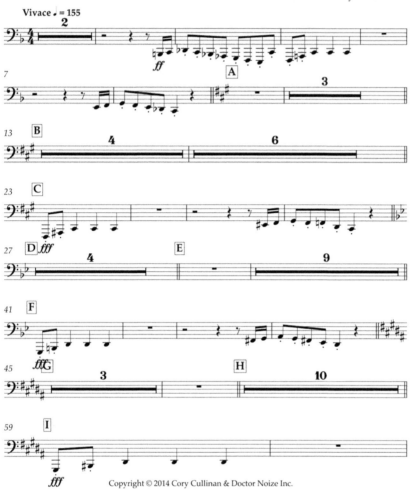

Figure 7.11 Excerpt

Despite the fact that I had been down in my production facility Reach Studios until 1am every night for months toiling away at the digital scores, awakening at 6am to get back to work; despite the fact that my wife was convinced I had become like Richard Dreyfus in *Close Encounters of the Third Kind*, Captain Ahab in Moby Dick, or Mozart with far less talent writing his Requiem in *Amadeus*; despite the fact that I had raised and accepted over $100,000 from generous music-loving fans and donors who wanted to see this work completed; and despite the fact that I had planned production and the recording phase over a year in advance … I boarded a plane in Denver scheduled to arrive in Prague about 15 hours later sleep-deprived and with four critical full scenes still yet to compose and arrange for the orchestra I had just committed $45,000 to record the project. The music was in my head, but it was not in score form

I worked the entire set of flights on the scores. I quickly realized it was far more laborious work than at Reach Studios because instead of my 88-key weighted keyboard to enter notes into the score, I had to grab and place the notes by hand with my trackpad. Nothing to do at that point but work harder. I did not have noise-cancelling headphones, which was brutal. (I have never boarded a flight without noise-canceling headphones since.) I found a plug somewhere else on the plane and would take nap breaks to plug my computer in for a while. (Thankfully nobody can leave the plane halfway over the ocean if they steal your stuff, but still a desperate and dumb move.) When I landed, I had *two* of the four final scenes banged out – conductor's score and parts. But two still remained, and I was so tired I could hardly think

38 - Phineas McBoof Crashes The Symphony

from *Phineas McBoof Crashes The Symphony*

Mama

words & music by Doctor Noize

Figure 7.12 Excerpt

38 - Phineas McBoof Crashes The Symphony

Violoncello

from *Phineas McBoof Crashes The Symphony*

words & music by Doctor Noize

Figure 7.13 Excerpt

I got to my hotel, emailed the latest scores to the orchestra's copyist, and tried to sleep. But I couldn't, knowing I had not completed the scores but had accepted everyone's money for the project. So, I got up and finished one more that night instead of sleeping at all. When I showed up that first day with the City of Prague Philharmonic Orchestra, I probably looked like a corpse. Thankfully, I had flown my favorite conductor, Kyle Pickett, in with me. Kyle and I had known each other forever; we were Stanford music students together, and he had conducted our tour of orchestral shows. He knew much of the music already, and I trusted that he understood my instincts if I was brain-dead at the recording session.

It was comforting that Kyle was there, because I had never met anyone in Prague before that day, and I was pretty sure the orchestra's musical director James Fitzpatrick thought I was an idiot American because I had called the Czech Republic "Czechoslovakia" numerous times in pre-production email exchanges, even though the country hadn't been called that in 20 years. (In James' defense, his judgment of me was only because I was an idiot American who had called the Czech Republic "Czechoslovakia" numerous times in pre-production email exchanges, even though the country hadn't been called that in 20 years.)

The first hour was terrifying, as big budget sessions can be, because of something every veteran producer knows but I was too tired to emotionally chill about: The orchestra sounded pretty terrible on the first run-through of the first few scenes. This is because recording orchestras generally don't rehearse anything in advance except the ones the music director (James in this case) tells them are particularly challenging. So, on the first run through, you get panic attacks as you think: "Did I just waste $45,000 of my funders' money??? This is my only shot to record this!" With a great orchestra like the one in Prague, however, by the third take they sound amazing. Phew. This realization happens quickly, as you don't record multiple minutes all the way through for such sessions; the orchestra rehearses and records bite-size chunks of a minute here and a minute there, and then we crossfade the best takes together digitally. This could be the subject of an entirely different chapter, but back to the session …

Another strange experience is that you are so used to the sound of the digital version your software has been playing back to you while composing, that the real

Figure 7.14 Back at the Studio

version sometimes sounds weird at first. Good … but *different*. You learn to ignore this quickly, realizing that just because the synth version is familiar doesn't mean it's better. At all. You remind yourself there's a reason you flew across the globe to record with the real thing. Now, when I listen to the software's version compared to the final luminous orchestral recording, it's hilarious that I even had this experience. But our mind loves what's familiar

After the first day of recording and no sleep the night before, I stumbled back to the hotel, did a workout and ate, and crashed for 13 hours of straight sleep. The next day, after more sleep than I'd had in months, was delightful. After two successful full days of recording, the good news is that the orchestra sounded amazing on the re-cording. The players were stellar, James ran an amazingly tight ship, and the recording engineer and Control Room engineer Jan Holzer in Prague was amazeballs. (Totally a word.) The bad news was that we were several hours behind schedule, making it un-likely that we'd finish recording all the scenes in the half-day session remaining to-morrow; I was out of money and could not hire the orchestra for another half day; and there was the pesky issue that I still had not completed the *final two scenes* completing the storyline and message of the work.

My only solution for this – as with most things – was to simply work with positive energy and do what I could do. We resolved to work faster, and record fewer takes the next day, which would obviously reduce the quality of the performances but give us a shot to record everything. And I pledged to stay up all night – for the second night in three nights – to finish the final scores. Here we go again …

Figure 7.15 In the Session

At 7am before an 8am to noon session, I completed the full score of the last two scenes. But ... I had not even started to extract the 30 parts for the different instruments. I was so close, but totally screwed. I sent the Sibelius score files to James and his copyist, told him the status, and grabbed a quick shower in the weird European shower in my hotel room (who builds a shower without a shower door?) and breakfast. Then I stumbled down the street to the orchestra's recording hall, right past the actual theater where Mozart's *Don Giovanni* premiered, an inspirational reminder that I was in a city steeped in music history, yet was probably going to die of this production, much like Mozart died from his *Requiem*, except that mine was an opera for kids about an evil bunny, and Mozart was producing one of the greatest works in history. Whatever. He wrote the overture for *Don Giovanni* the night before the premiere; at least we had that in common.

And this, dear readers, is the moment where the brilliant partnership of the human spirit and modern digital orchestration saves the day. When I arrived, I was told two things: 1. James (who I had thought didn't really like me) had decided that what we were doing – bringing a new operatic work for kids teaching them to love and understand the orchestra – was so good and valuable that he had personally ponied up eight grand to hire the orchestra for another four hours in the afternoon so we could finish the recording at the quality level its music and purpose deserved; and 2. James had brought his copyist into the studio and promised this amazing man would open up the score files of my final two scenes, extract all the parts for the players, and get both the score and parts onto their podiums by the new afternoon session. It was a testament to the previously-unheard-of collaboration that digital orchestrating software allows. If you have the same software and the same knowledge base, you can collaborate magnificently – even though the City of Prague's copyist and I literally didn't even speak the same verbal language.

Figure 7.16 Additional Parts

(Pause for me to cry tears of gratitude to James, his orchestra, his copyist, and the makers of Sibelius as I did that morning when I arrived.... Okay, back to the book ...)

That afternoon, after our lunch break, I was more exhausted and grateful than I had ever been. I listened to the final scene produced by 65 musicians, a talented conductor, an accomplished recording engineer, a generous and purposeful music director, and a brain-dead composer. It was the scene I wrote between 3am and 7am that very morning, and despite that, for some strange confluence of reasons, it is perhaps the most beautiful and profound music of the whole two hours and fifteen minutes of the work. It summarizes the purpose of the work brilliantly, and sounds like something beautiful and meaningful that a huge group of people collaboratively pulled off through toil, commitment and trouble. It was amazing, and a testament to the combined power of talent and technology. As Leonard Bernstein famously quipped: "To achieve great things, two things are needed; a plan, and not quite enough time."

Ultimately, my concerns and everyone's commitment – from funders to software programmers to musicians – were rewarded. The recording was extremely well-received – it's the only recording I've ever made that got 100 published reviews, and literally all of them were glowing. It was universally lauded as an old-school recording for kids reminiscent of *Peter & The Wolf* or Britten's *Young Person's Guide To The Orchestra*, *School Library Journal* named it an Essential Recording, it's now a part of the collection of libraries and homes throughout America, and now kids, parents and teachers can take home a copy of an audiophile recording of the work after they see the orchestral shows – or hear it without even going to a show at all. And while I'm proud to feel that its old-school, production-quality street cred is justified – even the idea of a composer writing a full two-act opera for normal newbies who don't drive expensive cars to the opera is so audaciously outdated today as to be a fresh concept – the old-school recording of *Phineas McBoof Crashes the Symphony* would never have materialized without full-scale digital orchestrating and recording technology.

Figure 7.17 Excerpt

Figure 7.18 Excerpt

Figure 7.19 Excerpt

Chapter 8

An outlook on future digital musical instruments

Introduction: the concept of NIME (new interfaces for musical expression)

As early as 1959, when the great 20th century French composer Edgard Varèse first brought computer music to the public eye, he stated that music as an art-science "has been at the forefront of many advances in the field of science and technology" (Risset, 2004). This fact has not changed since. As traditional acoustic and analog electronic instruments reached the limitation of providing composers new perspectives and freedom to create music in the early to mid-20th century, digital musical instruments (DMI) started to take over. As new technologies have appeared, inventors and musicians have been driven to apply new concepts and ideas to improve musical instruments or create entirely new means of controlling and generating musical sounds and interactive performance systems. Since the second half of the 20th century and the explosion of digital technology, the computer has become an essential tool for creating, manipulating, and analyzing sound. Its precision, high computational performance, and capacity to generate almost any form of audio signal make it a compelling platform for musical expression and experimentation. Moreover, the availability of many new forms of human interfaces, ranging from camera trackers to tactile displays, offers an infinite spectrum of possibilities for developing new techniques that map human gestures to manipulate sophisticated multisensory interactions and multimedia events.

In 2001, at the ACM Conference on Human Factors in Computing Systems (CHI), a workshop called NIME, with a concert and demo sessions, opened up a new chapter of DMI design. According to the editors of the book of *A NIME Reader*, NIME is defined as:

N = New, Novel ...

I = Interfaces, Instruments ...

M = Musical, Multimedial ...

E = Expression, Exploration ...

(Jensenius and Lyons, 2017)

Over the past 18 years, NIME has become a major conference in the fields of computer music, music technology, and DMI design and development. Researchers, developers, technologists, designers, and artists from all over the world work as a community and aim to contribute to the goal of making DMIs and Human-Computer Interaction (HCI) interfaces that can advance novel musical expressions and human repertoire. The community has been expanded from research in "interfaces" to a whole wild range of research on DMIs and musical interactions (Tanaka et al., 2010). Today, the concept of "NIME" has been expanded and can be interchangeable with state-of-art novel DMIs. People who design, make, and play these DMIs are called "NIMEers." The most prestigious NIME-making competition is the Guthman Music Instrument Competition,[1] which was founded in 2009 at Georgia Tech. The judges of the competition have included pioneer musicians such as Jordon Rudess, Pamela Z, and Richard Devine, who use NIMEs as their primary instruments, as well as many pioneer music technologists and scholars, such as Tod Machover and Joe Paradiso from the MIT Media Lab, Perry Cook from Princeton University, and Ge Wang from Stanford University. For a decade, this competition has been helping broaden the boundaries of NIME-making and introduce the NIME concept to the general public.

Two main approaches to DMI design

Although there are numerous ways of cataloging DMI design based on different technologies and research focuses, the main two approaches in DMI design are 1. Augmented DMIs, which use digital and sensing technology to extend/augment existing acoustic music instruments and derive information from the acoustic instruments; and 2. Innovative DMIs, which are completely novel digital music instruments with novel input, control, and mapping strategies that do not look or sound similar to any existing acoustic instruments. Major research has been done into these two main approaches (Roads, 1996; Chadabe, 1997; Eduardo Reck and Wanderley, 2006). Due to limited space, we will look into a few most recent examples to further explain these approaches.

Figure 8.1 Thorn with the *Transference* augmented violin (Photo courtesy of Thorn)

Augmented DMIs: augmenting traditional acoustic instruments

This kind of DMI can usually produce acoustic sounds stand-alone, through a physical instrument body and its acoustic mechanism. Augmented by extra hardware, sensors, and a software digital audio signal processing system, these DMIs can produce many different sounds, timbres, and novel musical expressions beyond the capacity of a traditional acoustic instrument. Augmented DMIs use existing gestural/movement input and control strategies that are similar to the way of playing the un-augmented traditional acoustic instrument, while the mapping strategies can vary wildly. One of the most recent examples of this kind is *Transference*, which is an augmented acoustic violin with a microphone input, an electromyographic (EMG) sensor and a custom glove controller to track hand and arm movements, as well as a customized ergonomic shoulder rest embedded with voice coils for haptic feedback coupled to digital audio output. Through the software system's digital signal processing, these gestures "actuate and perturb streams of computationally transmuted audio" (Thorn, 2019). Figure 8.1 shows the inventor, Thorn, playing his *Transference*

NIMEers not only have been rethinking and reinventing traditional Western acoustic instruments' possibilities but also exploring the new life of ancient Eastern instruments. For example, *SlowQin* resembles its predecessor *Guqin,* an instrument with more than two thousand years of history. *Slowqin* is fitted with a B Band piezo pickup under the bridge. A wide range of sensors is connected to an Arduino Mega combined with a JeeLink for wireless connection to computer software. The sensory body of the *SlowQin* is comprised of seven switches, four pushbuttons, eight potentiometers, a light sensor, two pressure sensors, and a long slide potentiometer with seven synthetic silk strings. In the current implementation, most of these input control data are mapped to multiple manipulations for real-time sound processing/synthesis, including polyphonic pitch, pentatonic shifts, granular synthesis, and modulation layering, applying complex sound effects on acoustic inputs and multi-modes live looping. These facilities rely on the software, or the "digital brain" of the *Slowqin*, which is a standalone SuperCollider[2] application, contributed by Hoelzl and Hildebrand, collaborating with the Chinese-German musician Echo Ho (Ho et al., 2019). Figure 8.2 shows a *Slowqin*.

Figure 8.2 Echo Ho playing the *Slowqin* in a live performance (Photo courtesy of Ho)

The main advantage of such DMIs is that they provide many new possibilities for musical performance, such as rich layers of soundscape, timbres, and musical expressions (Lahdeoja et al., 2009). The disadvantage of such DMIs is that they create difficulties for the performers to master and handle the extra effects and functionalities of the augmented instrument (Cook, 2001), thus preventing the performer from developing virtuosity (Wessel and Wright 2002). This is because the process of playing these DMIs often need the performer's extra level(s) of cognitive capacity, which is often beyond the cognitive capacity of just playing the original acoustic instrument. Specifically, this causes more confusion and sensorimotor-coupling mismatch for the experienced performers of the original acoustic instrument because they have already developed neuroplasticity in their brain for playing the original instrument (Wan and Schlaug, 2010), and the sensorimotor strategies that they use to engage with the original instrument is considerably different with the augmented DMI (Hrueger, 2014).

These DMIs should be treated as new instruments, on which performers develop virtuosity through intensive practices like electronic synthesizers and organs are not pianos. On the other hand, a piano is not a harpsichord either. Although a familiar control panel – the keyboard – may be helpful when a performer first starts practicing the instrument, she or he would need to work on this specific new instrument to develop virtuosity from this point.

Innovative DMIs: novel input, control, and mapping strategies

The ancestor of innovative DMIs: the theremin

One of the most well-known pioneers of the innovative DMI is the theremin, invented and patented in the 1920s by Russian engineer Léon Theremin (Glinsky, 2000). The theremin *was* the first touchless musical instrument in human history. The control mechanism of the theremin is a performer's hands contactless movement, which remotely affects the capacitance of the instrument's electronic circuits through two antennas, thus controlling an electronic oscillator's pitch and volume variations (Skeldon et al. 1998). Figure 8.3 shows a theremin.

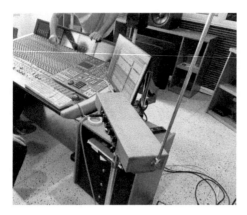

Figure 8.3 Theremin

Since the theremin was an analog electronic music instrument that only contained analog circuits and the only sound source was an electronic oscillator, the gesture-to-sound mapping strategy of the theremin was a "one-to-one" mapping strategy. This means one gesture/movement of a performer produces one sound result. Although the theremin used a simple "one-to-one" mapping strategy, its rich variations in sonic expressions caused by subtle hand-arm-finger movements have fascinated many musicians since it was born. Composers from the early 1940s to nowadays such as Miklós Rózs, Bernard Herrmann, and Justin Hurwitz composed many pieces and film scoring music for this instrument; performers around the world also have been playing this instrument with great enthusiasm.

MIT graduate Xiao and her research team have taken the theremin to a new level in the digital domain. Their DMI *T-Voks* augments the theremin to control *Voks*. *Voks* is a vocal synthesizer that allows for real-time pitch, time scaling, vocal effort modification, and syllable sequencing for pre-recorded voice utterances. They use a theremin's frequency antenna to modify the output pitch of the target utterance while the amplitude antenna controls not only volume as usual, but also voice quality and vocal effort. Syllabic sequencing is handled by an additional binary controller (a pressure sensor) that is attached to the player's volume control hand (Xiao et al., 2019). Figure 8.4 shows Xiao playing the *T-Voks* and the augmented theremin

(a)   (b)

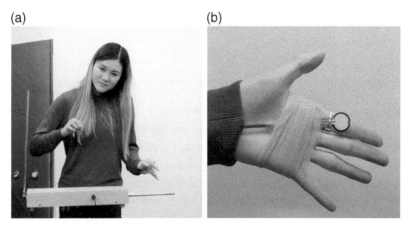

Figure 8.4 Xiao demonstrating *T-voks* and the augmented sensing technology (Photo courtesy of Xiao)

Figure 8.5 Spinning a Prayer Wheel (courtesy of Olivier Adam at www.dhar-maeye.com)

For mapping strategies, there are also "one-to-many" (which means one gesture/movement of a performer produces many sounds results) and "many-to-many" gesture-to-sound mappings for DMI design. Mapping, data input and control strategies are the core parts of the DMI design process (Claude and Wanderley, 2000). We will further discuss this design process and concept in the following two subsections.

An innovative DMI: the Tibetan Singing Prayer Wheel

To give the reader a clear idea of innovative DMIs, let's put an innovative DMI named the Tibetan Singing Prayer Wheel (*TSPW*) under the microscope. *TSPW* is a hybrid of a Tibetan spiritual tool called "prayer wheel," a traditional musical instrument called "Tibetan singing bowl," and a novel 3-dimensional sensor-based human-computer interface that captures hand-arm motions to enrich musical expressions. Inspired by the Tibetan singing bowl, prayer wheel, and the shared circular movement playing these instruments, *TSPW* utilizes the "spinning the wheel" hand motions to perform virtual Tibetan singing bowls and process human voice in real-time. It combines sounds and gestures, along with voice modulation, physical modeling and granular synthesis, using electronic circuits, sensors, and customized software (Wu et al. 2015). Physical modeling and granular synthesis is explained further on pp. 246–249. Since the prayer wheel is not a musical instrument, taking advantage of the prayer wheel's form, physicality, and playability to make music and manipulate human voice is an innovative DMI design approach. Figure 8.5 shows the operation of a prayer wheel

All the electronics fit within the cylinder of the handheld TSPW, which is about 1.5" radius and 1.5" height, leaving room for the rod running through the middle. Rotational speed is measured using a four-pole circular magnet held in place around the threaded center rod and a hall-effect sensor glued to the top of the wheel. The fact that the weight of a *TSPW* is less than a pound makes it very portable.

As for the input strategy, the performer gives the system three input methods: vocals via a microphone; spinning and intuitive gestures from a sensor-augmented prayer wheel; and button presses on a four-button RF transmitter to toggle different combinations of sound processing layers. These inputs activate a virtual digital singing bowl, real-time granular synthesis, and voice processing. The speed of the spinning controls the size of the grains of the vocal granulation – faster spinning results in a more intense granular effect and vice versa; the up-and-down movement of the wheel controls the pitch-shift of the vocal processing and the virtual singing bowl – higher hand gestures result in higher pitches and vice versa. There is also a pressure sensor on the wheel to simulate the sound effect of striking a singing bowl when pressing the sensor. As for the control strategy, controller units include the electronic augmented prayer wheel, the human vocal apparatus and microphone, hand-arm gestures, and the transmitter key with customized circuit design. As for the mapping strategy, it is a "one-to-many" mapping: using the same circular/spinning gestures to control multiple sound-making layers enriches the vocalist's musical expression in a simple way, whereas the button controller provides "one-to-many" mapping options and allows the performer to flexibly trigger eight different combinations of sound processing techniques during a performance. Figure 8.6 demonstrates *TSPW*'s system architecture.

For the sounds, the customized software consists of the physical modeling of a Tibetan singing bowl to simulate the sound of rubbing and striking the bowl; a granular vocal effect; and a modal reverberator that resonates with different harmonic partials of the voice input, as if the mallet used to play the bowl were injecting a singing voice into the bowl. The first and last effects were implemented in music programing language Faust[3] and then ported to Pure Data,[4] and the granular vocal effect was implemented directly in Pure Data through a laptop. These two music programming languages will be explored on pp. 245–248. Figure 8.7 shows the entire *TSPW* system with all its hardware and software components.

Since it was built, *TSPW* has been used in many concerts as well as for public engagement such as interactive installations and exhibitions From the observations of these applications and events, playing *TSPW* provides an intuitive and intimate way to connect the user's physical movements to her/his sound experience, as they are making, and aesthetically appreciating, perceiving, and enjoying sound with her/his own bodily activities (Wu, 2018). The DMI focuses on incorporating enactive and auditory feedback in a novel way, as well as providing the user with an embodied sonic experience.

DMI system architecture: input, control, and mapping strategies

As shown in Figure 8.6 and discussed in detail on p. 239, a typical system architecture of DMIs always contains a control mechanism, which is controlled by or can interact with the performer. It is not uncommon that the performer simultaneously controls multiple controllers to achieve an ideal musical outcome. Sensing technology and hardware/circuit design for human-computer interaction often are implemented to build the controllers. This control mechanism can take in gestural/body movement, sonic, visual, optical, heat, or any kind of raw data as input.

Through a data optimization and scaling algorithm, the raw input data can be translated to useful information and can be applied to map into audio processing. Nowadays, mapping mechanisms and sound machine implementations are usually a hardware-software hybrid. Sometimes, designers develop completely software-based mapping systems to lower the cost of computation and allow for faster digital audio signal processing.

Moreover, these innovative DMIs do not depend on any specific physical/acoustic mechanism, but use integrated circuits, sensors, digital computing, and analog-digital converters to produce sounds. These sound generation methods of innovative DMIs set the designers and music technologists' minds free, as the DMIs can be made into any shape or form, by any material, visible or invisible to the audience. Although electric sound generation methods, theory, and implementations of digital audio are out of the scope of this chapter, readers can find detailed information in Miranda's (2012) book as well as in Boulanger and Lazzarini's (2010) book.

Nostalgia and mutation: modular synths and virtual modular synths

The "West Coast" and "East Coast"

The gigantic Telharmonium is the earliest synthesizer and can be traced back to 1906. However, all the synthesizers before the 1960s were very expansive, large, heavy, and difficult to operate. Therefore, they were only used in privileged high-end audio studios and laboratories such as the West German Radio (WDR) and Columbia- Princeton Electronic Music Center (CPEMC) (Roads, 2018). Due to the massive availability of $0.25/piece cheap transistors in the 1960s, the commercial and relatively affordable analog modular synthesizers were first available in 1963. During that year, modular synthesizer pioneers Robert Moog from New York City and Don Buchla from California both established their modular synth companies, without knowing each other. Since then, the unique sounds of Moog synthesizers have been called the "East Coast" style and the equally unique sounds of Buchla synthesizers have been called the "West Coast" style (Pinch, 2016). The other iconic modular synthesizers, invented in the late 1960 and in the 1970s, include the ARP, invented by Alan R. Pealman, and the EMS by Dr. Peter Zinovieff.

The core technology of analog modular synthesizers is the variety of methods in controlling incoming and inter-between voltage variations. Early analog modular synthesizers often had separated voltage control units, such as sound oscillators, amplifiers, filters, envelopes, and later sequencers (Pinch and Trocco, 1998). Moog was the first engineer who built a physical ADSR envelope shaper. These units could be connected by a patch board and cables thus changing the frequencies, dynamics, and timbres. Without a keyboard controller and with many potentiometer knobs, switches, patch board, and sometimes also an oscilloscope, these control units and modules looked much more like scientific equipment than musical instruments. Therefore, Robert Moog adapted the conventional musical keyboard as Moog Synthesizers' main controller.

The most influential promoter of Moog's keyboard-attached synthesizers was the transgender musician Wendy Carlos. She first started using the Moog synthesizer 900 series around the year 1967. Her multi-Grammy-award winning album series, *Switched-on Bach I and II*, were published in the late 1960s and mid-1970s by Columbia Records and have received significant commercial and artistic success (Reed, 1985).

On the other side, as an experimental musician, Don Buchla went much more radical in controller design, as he did not want the underlying implications of the traditional twelve-tonal musical structure of the keyboard to limit his and other experimental musicians' imaginations (Aikin's *Keyboard* interview with Buchla,[5] 1982, cited by Pinch and Trocco, 1998). Because of his distaste of keyboard controllers, Buchla designed his famous alternative tactile controller *Thunder* in 1989, just a few years after the MIDI standard was established (Rich, 1990). Just like the system architectures of Buchla's modular synth systems have been continued to be cloned, *Thunder*'s mutational variations such as the Sensel Morph's Buchla Thunder Overlay[6] (Lui-Delange et al. 2018) and the Thunder control modules such as Buchla 222e, 223e, and Easel-K[7] series are being widely used in today's modular synth field.

Arguably, one of the most accomplished musicians who uses Buchla synthesizers as their primary instrument is Suzanne Ciani. Ciani was also the first and one of very few women working in sound design in the early days of the 1970s. In contrast with Wendy Carlos, who played Moog's keyboard synthesizers and received many album contracts at the beginning of her career, Ciani's music-making using a non-keyboard Buchla 200 synth and wild creation of unfamiliar electronic sounds instead of a piano-alike or organ-alike sound on a keyboard was not understood until the recent years. Ciani recalled the old days that she could not even nail down a single album deal because her music was "too weird" and her instrument looked like alien equipment from the moon. She says,

> In the 60s, that new world was aborted to some degree. The idea of a synthesizer was taken over by cultural forces that didn't understand the potential: "Oh, you can synthesize the sound of a flute," "Can you make the sound of a violin?" [There was] this preoccupation with copying existing timbres. I thought it was a dead idea.
>
> (Interview by Hutchinson, 2017)

What Chiai described above contributes to one of the reasons why most of the analog modular synthesizers company went out of business in the 1980s when computer and digital technology are available to the world. John Chowning's patented FX synthesis-based Yamaha DX7 was the first digital synthesizer and it was sold more than 200,000 units because of its affordability and aligning with the idea of synthesizing existing instruments. Gradually, analog synthesizers became cumbersome dinosaurs and people lost their interest until the new millennium.

The modern Eurorack scene

Dieter Doepfer was the pioneer who designed the first Eurorack system A100 and launched it in 1995 (Bjorn and Meyer 2018). Doepfer successfully lowered the cost of modular synthesis by using only three vertical rack mounting units in height and making it more compact as well as implementing digital technology and computer algorithms. The 3.5 mm mono jacks were also different from the past. According to ModularGrid's[8] 2018 statistics, "Eurorack has become a dominant hardware modular synthesizer format, with over 7100 modules available from more than 316 different manufacturers" (Roads, 2018). For a detailed review of the recent ten years of the trends of modular synthesizers, readers can visit the modular synth online forum.[9] Figure 8.11 shows the Doepfer A-100 first Eurorack system.

So why are Eurorack modular synthesizers so popular in the digital era and still going strong? Are the software-based virtual machines using music programming languages and VST plugins in the DAW not good enough? What is missing? The answer to this question will give important hints as to the future of DMI design and development.

From the author's perspective of practicing on a series of modular synthesizers over the last six years, and within the historical context, the main charms of modular synthesizers, specifically, the Eurorack modular synthesizers, are the following:

1. the flexibility of putting different modules together and the freedom to choose what can be in the system. This is an intimate design process. It creates the musician's ownership of her/his own preferred musical instrument and the instrument can constantly be evolved by just changing one or a few modules or the sequence of patching cables;
2. compared to the "inbox" software-based music creation, it provides expressive gestural expressions, tactile feedback, and engaging interactions with the system; for example, patching the cables, playing with the sensor-based control units, (e.g., tuning the knobs and sliders) creates a better-embodied experience and a strong sense of "playfulness" than the pure software-based virtual machines so the musician feels intimately connected to the instrument;
3. the unpredictability and immediacy of the system give life to the machine and excitement to the musician. As a hybrid analog-digital voltage-controlled system, Eurorack systems inherit the unpredictable and direct natures of their ancestors. Sometimes, a musician can only know what sonic behavior would come out of this "little monster" after she or he tunes 1/5 of a knob or makes a subtle change of the patching and this unexpected sound slaps your face and shocks you right there at the moment, which makes the music-making experience very entertaining. Each interaction with the system is like an adventure and an unknown journey;
4. empowered by digital computing and digital audio technologies, the cost of Eurorack modular synthesizers are cheaper and the weight is much lighter than the analog synth from the old days. With adequate computer memory and faster microprocessors, musicians now also can compute and store their sequencers, sounds, and music creation and bring them back to life anytime;

5. new hybrid Eurorack modules also enrich the current modular synth's scene "via the emulation of process and methodologies of the past such as tape emulation, time stretching and granular synthesis approaches" so musicians can revisit and explore those vintage compositional methods with a historical context and find the aesthetic value from it (Connor, n. d.);

6. the open-source, hackable software/hardware implementations and the well-documented shared system structures available online make Eurorack the infinite playground of music and technology lovers. Any music geek can clone or build her/his own modules based on this hands-on information. This is also the reason why there are so many new Eurorack modules around the world. Indeed, the "do it yourself" philosophy attracts musicians, technologists, and hobbyists to invent new modules, thus enriching the catalog of Eurorack design and implementation.

Virtual modular synth patching

Although compared to the old-times analog synthesizers, Eurorack modules are much affordable, most of them are still in the price range of $100–500/module. Meanwhile, besides the main oscillator, filter, envelope, modulation, and sequencer modules, users need to purchase rack, case, utility modules, and many other small but necessary parts to connect the modules and make them a complete musical system. Because of this, a one-vertical-unit Eurorack system can easily cost more than $1,500, not to mention the time and labor one has to put into the soldering and research on how to do this by oneself. Therefore, a lot of beginners, hackers, and geeks prefer a "jumpstart" solution that is more affordable, or ideally, completely free, so they can get their hands dirty and start to learn and make some noises.

This is where the virtual modules and virtual racks shine. Currently, the one free and open-source cross-platform virtual Eurorack Digital Audio Workstation (DAW) is VCV Rack. It provides a virtual rack and many free virtual Eurorack modules that users can download and play with. As a free and open-sourced framework, VCV Rack has some advantages over other similar commercial products. First, the human-computer interaction design of this virtual Eurorack DAW is intuitive, transparent, and easy-to-use, and all the virtual modules' manuals are well documented in the product and at Github,[10] so developers can improve the DAW as well as making new modules and contribute to the VCV Rack community in real-time.

In addition, VCV Rack and Mutable Instruments collaborated on a plug-in called Audible Instruments, which includes most of the very popular modules that Mutable Instruments released over the years, such as the *Braids* (a sound generator), *Clouds* (a granular synthesizer), and *Rings* (ring modulation) and now they are completely free for everyone. This way, people can explore these modules before they commit to purchasing one. Meanwhile, for those very popular but discontinued modules such as *Braids* and *Clouds*, people can still have them at some point in their virtual rack if their real physical modules are broken. Lastly, open source also means that instead of waiting for upgrades of commercial products and paying for each version of the upgrades, users can expect new upgrades every day, every moment, for free.

Much more could be written about this fascinating modular synth, Eurorack, and the virtual Eurorack landscape, such as the Make Noise,[11] WMD,[12] 4ms,[13] Reaktor6[14] and experimental physical modeling virtual modules *Aalto* and *Kaivo*.[15] In fact, ModularGrid provides a detailed "Top100 modules list"[16] based on users' feedback.

In brief, the historical development and innovations in modular synthesizers and virtual modular synthesizers have made a significant impact on DMI designs, musicians' compositional decisions and aesthetic choices, as well as how musicians can intimately interact with their instruments and the audience to express their musicality. Because of the reasons stated on pp. 243–244, modular synthesizers will continue to be popular and well-received by even a broader audience in the future.

Performing music on-the-fly: music programming-based systems (virtual machines)

Since home computer became more powerful and computer storage became cheaper in the late 1990s, Digital Music Workstations (DAW) and VST plugins have become the main tools for many professional and amateur musicians to record, edit, and mix music "in the box." Although the DAW has been a great tool for fixed music production, most DAWs share the same VST plugins, so sound creation and processing are somewhat constrained. For many experimental musicians who wanted more freedom and flexibility to make unique sounds on-the-fly and create their own musical tools, customized music programming-based systems have been the best options.

Although maybe unfamiliar to many readers, music programming-based systems have already had more than 60 years' history. In 1957, the father of computer music, Max Mathews, wrote the first music programming language named "MUSIC I" for digital audio generation and processing using direct synthesis in a computer (Roads, 1980). Since then, many music programming languages such as Csound (Vercoe, 1986), Kyma (Scaletti, 1989), Pure Data (Puckette, 1997), MAX/MSP (Zicarelli, 1997), Supercollider (McCartney, 2002), FAUST (Gaudrain and Orlarey, 2003), and ChucK (Wang and Cook, 2003) have been developed and widely used. Using these frameworks allow musicians to build their own "virtual machines" to do real-time sound synthesis, algorithmic composition, and interactive controls in live performances. No off-line rendering time is needed – everything is made on-the-fly with little latency and high computational and timing precision.

Many music programming languages and coding environments are open-source and free, such as Csound, Supercollider, Pure Date, FAUST, and ChucK (we will further mention ChucK on pp. 251–253). On the other hand, MAX/MSP and Kyma are commercial products. Although almost all of these popular music programming languages can be used for real-time multimedia installation and live performance, customized DMI design, and physical computing, each of them has its unique strength and capacity and suggests its own way of music-making. For example, Csound, Supercollider, and ChucK both support the traditional coding method of writing high-level code in lines – just like the old-school computer programmers writing normal code; whereas PureData, MAX/MSP, and Kyma use graphical programing methods to link sound generators to different control and processing modules – just like the old days of using modular synthesizers and patching cables.

Furthermore, different programming languages behave in their own ways and also "sound" differently. For instance, although both the two music programming languages were invented more than 30 years ago, Csound was first designed to be note-oriented and completely standalone with its software environment to generate sounds that are distinguished from other music programming languages; whereas Kyma has always been focusing on an abstract sound-design approach and making powerful digital signal processing (DSP) hardware systems for generating high quality sounds in real-time. Both these languages enable a range of their unique musical vocabularies and ways of communication – just like you would not confuse the Chinese language with English or Italian.

Since *TSPW*, described in pp. 239–241, was written in two different music programming languages (PureData and FAUST), we will use *TSPW* to explain how these programming languages work in this DMI design scenario. For the sound generation part, a physical model of a Tibetan singing bowl was computer simulated using waveguide synthesis (Smith, 1992) via the Faust-STK toolkit (Michon and Smith, 2011). Before waveguide synthesis, researchers were using finite-difference schemes (numerical integration of partial differential equations), or modal synthesis models to simulate acoustic instruments. There was no way to interact with the virtual sound wave propagation medium (e.g., to truly simulate playing an instrument accurately). Waveguide synthesis provides extreme efficiency of computing and makes real-time wave-propagation, and interaction with it, feasible in real-time. It enabled the Yamaha VL series of virtual acoustic synthesizers to appear in 1994. Figure 8.6 shows a photo of Yamaha VL and its technology inventor Professor Julius O. Smith at Stanford University's Center for Computer Research in Music and Acoustics (CCRMA)

The great advantage of waveguide synthesis or physical modeling is that it offers extremely authentic synthesized sounds that sound just like what you would hear when playing the original acoustic instruments. Waveguide synthesis can simulate the

Figure 8.6 The First Physical Modeling Synthesizer Yamaha VL (Photo courtesy of Julius O. Smith)

force, speed, and direction of plugging a string, hitting a drum, rubbing a singing bowl, or you-name-it methods of interacting with an acoustic instrument, and then real-time render a realistic sound of this physical interaction. Indeed, it was very expensive to implement this highly computationally intensive technique in the 1990s. The Yamaha VL1 cost $10,000 on its launch in 1994. Therefore, physical modeling was not widely applied when computer technology was not enough to support it in an affordable way. This is no longer a problem nowadays. Physical modeling, especially waveguide synthesis, has become the state-of-the-art technique of synthesizing acoustic music instruments in the digital world.

With this technology background in mind, starting from 2002, Yann Orlarey and a few other researchers in France invented the FAUST music programming language, to generate real-time digital signal processing (DSP) applications and plugins for many other music programming languages such as ChucK, MAX/MSP, and PureData, from a high-level specification. FAUST is a DSP powerhouse and an audio plugins generator. Many physical modeling plugins can be much easily designed, written, and generated via FAUST and then imported to any other popular music programming environments for implementation and integration. The design of *TSPW* uses this method–the physical model is fully implemented in FAUST and was compiled as a Pure Data external plugin called "instrument" (shown in Figure 8.8). This "instrument" module then can be patched to a motion detector – Hall Effect sensor – to generate "rubbing a singing bowl" sounds when a performer is spinning a *TSPW*.

Although one part of the physical modeling implementation was written in FAUST music programming language and then imported to Pure Data (PD) as a plugin, *TSPW's* software development part and the hardware/software communication part were all written and integrated in PD. PD's author, Miller Smith Puckette, is also the original author of MAX/MSP, back in the 1980s. The title of "MAX/MSP" is actually dedicated to the father of computer music–Max Mathews (the part of "MAX), and Miller Smith Puckette (the part of "MSP" uses his initials), by the later MAX/MSP's main developer David Zicarelli and his company Cycling 74.[17] Therefore, PD and MAX/MSP share a considerable amount of common ground and architecture design philosophy.

The main differences between PD and MAX/MSP is that PD is completely free and open-source and thus attracting many computer musicians and developers; whereas MAX/MSP is a successful commercial product that offers a better (or fancier) graphical user interface (GUI) and arguably better-maintained patch examples that can be grabbed and used instantly without debugging or modification. The greatest advantage of using MAX/MSP is that it can be integrated with Ableton Live[18] to modify any Ableton Live's plugin settings or create a new plugin and hook the system to customized controllers and music machines. This way, it creates a much more powerful DAW that enables experimental music performances. This integration offers extra layers of freedom, flexibility, and creativity that a traditional DAW is lacking, thus pushing the boundaries of using DAWs as an undefined DMI in live performances.

The overview of the PD patches of *TSPW* is shown in Figure 8.13. On the left of Figure 8.14 shows a PD program making granular vocal processing effect, in which the human voice is cut into small "grains" and the granulated vocal fragments are played back while the singer is singing in real-time. The length of each granulated vocal fragment depends on the speed of the TSPW rotation, as explained on pp. 239–241.

While physical modeling is good at synthesizing acoustic music instruments in an authentic way, granular synthesis creates alien-like abstract and novel sounds. Since the new century, granular synthesis has become one of the most popular and chic synthesis techniques as the computer technology has grown powerful enough to granulate audio at a high sampling-rate and process this massive amount of data in real-time with ease. In the 1940s, the Nobel prize-winning physicist Dennis Gabor (1900–1970) first proposed the concept of decomposing sound into "acoustical quanta bounded by discrete units of time and frequency" (Roads, 2004). This was the first time a continuous sound was treated as discrete units. Gabor was the first person who built a machine to granulate sound into particles in the analog domain.

Greek musician, architect, and mathematician, Iannis Xenakis, extended this concept into his famous music theory in "formalized music" (Xenakis, 1971, 1992). However, it was too tedious for him to implement granular synthesis in the digital domain because of the limited computer technology. One of Xenakis' students, Curtis Roads, recalled that the way that Xenakis realized granular synthesis in the analog domain was to record tones on tape and then cut the tape with scissors into small pieces. Xenakis put all the pieces for one tone into an analog audio envelope shaper, so he had many envelopes with the countless bits of tape. He then generated a score. His assistant, Bernard Parmegiani, a French composer, was given the task of splicing together all the little bits of tape to realize Xenakis' score. When Parmegiani had lunch with Roads around 1995, he told Roads about how difficult and tedious this task was.

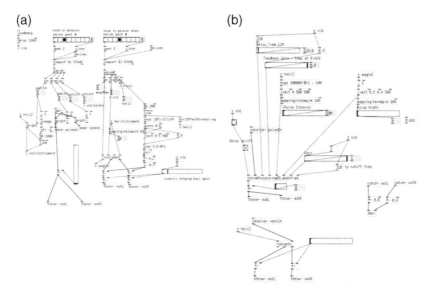

Figure 8.7 a & b Figure 14. shows hardware-soft communication patch (left), and the Tibetan Singing Bowl physical model real-time manipulation patch (middle) and the granular vocal processing effect for real-time processing the human voice (right).

Figure 8.8 A Distributed Electroacoustic Network Improvisation (Photo courtesy of CCRMA, Stanford)

Dr. Roads became the first person to implement granular synthesis in the digital domain, using a music programming language named "MUSIC V" – the later version of the first music programming language "MUSIC I." Back to 1974, computer technology was still very primitive. Roads wrote in detail in his *Microsound* book about this implementation process:

> Owing to storage limitations, my sound synthesis experiments were limited to a maximum of one minute of monaural sound at a sampling rate of 20 kHz. It took several days to produce a minute of sound, because of the large number of steps involved…. I typed each grain specification (frequency, amplitude, duration) onto a separate punched card. A stack of about eight hundred punched cards corresponded to the instrument and score for thirty seconds of granular sound."
>
> (Roads, 2004)

Without doubt, the scientific value of granular synthesis is to operate sound at a micro-level, deconstruct a continuous sound into tiny discrete particles, both in the time and frequency domain. These values offer aesthetic, practical, and scholarly novelties to the musical world. Therefore, with the progress of time and the development of science and technology, granular synthesis will continue to shine and will be more and more powerful in music-making. Moreover, as computing and web audio technology becomes more advanced, web browser-based music systems that enable complex real-time sound synthesis and processing (e.g. granular synthesis and physical modeling, etc.) will become more and more feasible and popular.

As discussed, the *TSPW* example demonstrates that music programming languages can be used as powerful tools to build completely novel real-time interactive DMIs, create unique sounds that no existing VST plugins would be able to offer, and implement new functionality and affordance to costumed hardware design. As long as the computational power and material science allow, we can build a DMI and make it sound like anything we want, and make it look like anything we can imagine, with maximum freedom, flexibility, and creativity. With the cheaper and faster digital computing technology developing in the new century, this approach will become more practical, applicable, and critical in future DMI design.

DMIs for social engagement

Music is a socio-cultural product. In a societal context, music is produced, reproduced, consumed, and institutionalized (Martin, 1995). In the new century, DMIs play a crucial role in this production, reproduction, consumption, and institutionalization process. Understanding how DMIs, as a meaningful vehicle, can increase one's degree of participation in society will help us better understand the social environment and its historical background. In this section, we will investigate this matter from three perspectives: community building, virtual reality, and gaming, as well as new media exploration.

Building communities: social music-making

Before the late 1870s, recording technology was not available. If people wanted to listen to music, the only thing they could do was to go to a concert. People dressed up, sat in the same room, and enjoyed the same unduplicatable social-musical moment. Arguably, people at that time also had more in-person interactions because the main entertainment after dinner was playing music together instead of watching television or checking their cell phones on social media. As times have changed, can music technology bring people closer together instead of drifting apart? Music technologists have been contemplating this question for a long time, and some novel solutions have been proposed in recent years.

Network concert technology

As we have all experienced at some point, jitter and packet loss often occur when using common real-time network communication applications such as Facetime, Skype, or Google Hangouts. These real-time network communications actually are not real-time. A few seconds' latencies usually are inevitable. This latency is a big issue for online music gigging and performing through network communications as musicians must synchronize their time in order to play music accurately together. Imagine, for example, a band in which the drummer has already started playing two seconds ago and on the other side, the guitarist starts two seconds late – how could they possibly make a decent performance?

Music technologists have been working on this issue for two decades. One of the state-of-the-art network music technologies is Jacktrip.[19] Jacktrip is an open-source audio network communication protocol for high-quality audio network performance over the internet (Cáceres and Chafe, 2010). The technology explores ways of connecting cultures and collaborative artistic partners over long distances. Jacktrip was first developed in the early 2000s. It is now available for Linux, Mac OSX, and Windows-based systems and the different versions can connect to any of the others. It can be used for multi-machine network performance over an Internet connection. Jacktrip supports any number of channels of bidirectional, high quality, low-latency, and uncompressed audio-signal streaming.

Musicians have increasingly started using this emerging technology to play tightly synchronized music with other musicians who are located in different cities or countries. Without paying significant transportation costs, people have exchanged musical ideas, rehearsed, recorded, and improvised together in different geological locations. This completely free music technology brings people closer as musicians now can gig anytime, anywhere, even they are living on different continents. We encourage readers to further explore this technology.[20]

During the most recent COVID-19 quarantine, this technology has become really helpful to many musicians as they were not able to gather in-person to play gigs. Many musicians have been trying to give remote concerts and there is a surge in interest in network concert technology. For example, Stanford University's "Quarantine Sessions Series" applied this technology and hosted a series of network concerts that remotely real-time connected musicians from six to ten different cities. The sessions were broadcast live with audio and video feeds from each site. Figure 8.10 shows a glimpse of the "Quarantine Sessions Concert #5."

Laptop orchestra

Generally speaking, any computer-based meta-instruments ensemble can be called a "laptop orchestra." Some references prefer use "laptop orchestra" if they are chamber music or classical music-based and at the same time made inside of ivory towers – the universities. The first laptop orchestra – PLOrk–Princeton Laptop Orchestra[21] – was founded by two professors with their two graduate students at Princeton University in 2005. One of the graduate students at that time, Ge Wang, is also the architect of the ChucK music programming language, which was developed as part of his Ph.D. dissertation. Wang later founded the SLOrk–Stanford Laptop Orchestra[22] when he started teaching at Stanford University in 2008. Both PLOrk and SLOrk use ChucK as their primary music programming language and have expanded the capacity of ChucK during the past decade.

In the same year of 2008, BLOrk–Boulder Laptop Orchestra was founded at the University of Colorado Boulder. Later on, CLOrk–Concordia Laptop Orchestra, CMLO–Carnegie Mellon Laptop Orchestra, HELO–Huddersfield Experimental Laptop Orchestra, and OLO–Oslo Laptop Orchestra were founded respectively by many other universities around the world. City-based laptop orchestras include BiLE–Birmingham Laptop Ensemble and MiLO–Milwaukee Laptop Orchestra. The list of laptop ensembles/orchestras is still expanding. Gradually, laptop ensemble/orchestra have become a major theme in music in higher education.

The main idea of laptop orchestra is to use computer music technology and novel DMIs to compose, improvise, and perform new music expressions in a group setting, thus building communities and bringing communities together. Laptop orchestra tools have included laptops, hemispherical speakers, MIDI controllers, circuit-bending, and NIMEs. More often than not, they also include traditional acoustic instruments, voice, cultural music, multimedia display, and using sensors to track and sonify body movement, etc. The laptop orchestra is a combination of everything that you can imagine of novel musical expressions. It also opens another door, lowering the barriers to many non-classically trained musicians who are technology savvy, can master creative skills, and are dreaming of performing at formal concert halls.

At the same time, the laptop orchestra also has become an alternative educational tool for K–12 and college students to study computer programming from an artistic approach. Indeed, although the computer logic and mathematical theory behind the theme are the same, to many students, it is a lot of fun to write code and at the same time listening to the music the code has made. Moreover, many music programming languages are high-level computer languages and contain modules to generate, filter, and process sounds. It is convenient and easy to look into the original code of each module to figure out how electric sounds are synthesized and how analog audio is digitized and processed by the computer. The laptop orchestra can also be a pedagogical tool for computer music.

The most important impact that laptop orchestra has brought to society is that it offers new ways of in-person communication and social lifestyle, and expands the concert repertoire. It invites the younger generations to participate in non-commercial, serious art music with state-of-the-art technology, so younger people have many other aesthetic choices besides just listening to or making popular music or traditional Western classical music. People now can go back to the concert hall, participate in the laptop orchestra, and enjoy the excitement, novelty, and human connections through innovative ways of music-making.

Social music-making mobile applications

There are numerous music-making mobile applications on the market. Many of these applications are designed to be connected to the users' social media accounts so once the music is made it can be shared. One of the easiest social music-making mobile application companies is Smule.[23] It carries many social music-making products such as Ocarina, Magic Piano, and AutoRap. These mobile applications are human-centered, easy-to-use, and very encouraging for laypersons to make music. The company's philosophy is that music is for everyone, for connecting people and bringing us closer.

For example, Smule's Ocarina mobile application (Wang, 2014) takes advantage of the iPhone's embedded sensors and turns the iPhone into a DMI that simulates an ancient ocarina wind instrument. Users can practice on this DMI with well-paced tutorials, step by step. Once the user feels satisfied with the practice, he or she can record the piece and share it with all the other users around the world, using a network connection to a suite of cloud-based services. Recordings of control data are tagged with GPS data and uploaded. Users can view, listen to, and share their music and display their geographic locations on a rendered Globe or World Listener interface within each application and on smule.com. Links to their music recordings are shareable on social-networking sites such as Facebook and Twitter or can be sent via email (Hamilton et al., 2011).

DMIs for virtual reality and multimedia

Virtual reality technology development can be traced back to the 1960s. However, because of the costs and limitations of computing and sensing technology before the new millennium, it was not until the year of 2010, when the first prototype of Oculus Rift was built, that the massive commercial application of this technology began. In the DMI world, VR DMI design is still in its infancy. Although virtual machines, software DMIs, and physical modeling simulated musical instruments are somewhat "virtual," they are not considered VR DMIs. Serafin et al. (2016) suggested VR DMIs must include a simulated immersive visual component delivered using VR headsets or a three-dimensional immersive visualization space. Besides visuals, these VRMIs can also include all the elements of multimodal immersion methods such as auditory, haptic, smell, and taste.

The applications of VR DMIs can be used in many fields such as multimodal spatial audio-visual productions, music education, and entertainment. For example, in the commercial world, VR audio DAWs such as Nuendo[24] and DearVR[25] can seamlessly work with many VR gadgets to create immersive multisensory experiences. Often, these VR productions made by VR audio DAWs are pre-rendered and non-real-time; whereas in the open-source world, things can be much more flexible with real-time human-computer interactions. For instance, audio system developers can integrate an audio engine written in music programming languages such as SuperCollider and Pure Data to the Unity game engine via a network communication protocol called Open Sound Control (Wright, 2005). On the other hand, audio engines that are developed in Max/MSP and Chunity (ChucK's special programming environment for Unity) can be integrated without relying on any network communication (Atherton and Wang, 2018). A few VR DMI design examples can be seen on YouTube, such as *Crosscale* (Cabral, 2015) for entertainment, *ChromaChord* (Fillwork, 2015) and DeepMind12 Augmented Reality[26] for electronic music-making, and *Teomirn* for music education (Tung and Schnieders, 2018).

Extending this approach to a more generalized approach, multimedia and multisensory DMIs are everywhere. For example, Thorn's *Transference* augmented violin that we have seen earlier in this chapter is a haptic DMIs that combines the multisensory experience of touch and hearing, coupled the tactile and audio feedback together, whereas most of the VR DMIs are audio-visual DMIs.

A recent example of audio-visual DMI design, *Embodied Sonic Meditation work3–Resonance of the Heart*, uses an infrared remote sensing device and touchless hand gestures to produce various sonic and visual results. An artificial neural network was implemented to track and estimate the performer's subtle hand gestures using the infrared sensing device's output. Six sound filtering techniques were implemented to simultaneously process audio based on the gesture. Selected ancient Buddhist Mudra hand gestures were mapped to seven novel 4-dimensional Buddhabrot[27] fractal deformations in real-time. This DMI was applied in both college teaching, electro-acoustic live performances, and public art installations. It connects Eastern philosophy to cognitive science and mindfulness practice. It augments multidimensional spaces, art forms, and human cognitive feedback (Wu and Ren, 2019). Figure 8.8 shows the audio-visual DMI *Resonance of the Heart* is in use and a series of its real-time interactive fractals visual display.

The pioneer of media theory, Marshall McLuhan, famously stated that media and technology extend the human body (McLuhan, 1964). The machines we built have augmented us, so we can see, hear, touch, taste, smell, and feel what we previously cannot. Since the 1990s, people have been contemplating the fundamental questions concerning the relationship between humans and technology in this "posthuman" era (Hayles, 2008). From computer-assisted algorithmic composition beginning in the 1950s to 1960s biofeedback experiments to VR DMIs and wearable interfaces to intersections between music and artificial intelligence, the musical tools and media technology that we have been using gradually become a natural part of us and change our collective behavior in the society. In other words, we have become augmented musical posthumans through the music technology that we have been experimenting with and inventing (Cecchetto, 2013). We will further discuss this in the following section.

Posthuman DMIs: biotech, informatics, and artificial intelligence

In 1960, the first paper on algorithmic music composing was published (Zaripov, 1960). This signified the era of "posthuman sounds." Since then, computer algorithms, biotechnology, informatics, and artificial intelligence (AI) have gradually shaped our aesthetic decisions and changed our role in music/art creation. Indeed, the arts is almost the "last bastion" of human beings. Just imagine, once we are replaced in this creative world, what could not be replaced? Where would we be able to exercise our human agency? Would our species be replaced too? The British mathematician Irving Good once made a famous statement that the creation of artificial intelligence would be humanity's last invention (Good, 1965). He also anticipated the eventual advent of superhuman intelligence, now known as the "intelligence explosion" or "technological singularity" (Shanahan, 2015). This hypothesis predicts the growth of technology will one day become uncontrollable and irreversible, highly likely resulting in the termination of human civilization.

A recent music video named *Automatica* by New Zealand musician, Nigel Stanford (2017), artistically narrates this hypothetical idea of AI destroying human civilization. In the video, KUKA arm-robots[28] first play multiple instruments with the human musician in a harmonic way then destroy everything in the end. In fact, having a robot play musical instruments is nothing new. About 250 years ago, invented by Jaquet-Droz[29] sometime between 1773–1775, a very early automaton named "the Musician" was already able to play a 24-key harmonium with its mechanic-engineered wood fingers. In the early 1980s, a humanoid robot named "Wabot" could read musical scores and play an electronic keyboard (Sugano and Kato, 1987). The recent "Waseda Talker" robot series even can mimic a human's vocal cords (Fukui et al., 2012). Maybe, soon, we will be able to see robots singing expressively. These early automata, and today's many digital-empowered robots, follow precise prerecorded trajectories or are real-time controlled by human operators to produce sounds. If any instrument or part of the system is moved, then the sound production fails completely as the robot is unable to visually capture the position. In other words, these robots simply reproduce exactly what the humans programming them tell them to.

Music researchers nowadays are focusing on creating AI assistants who can help and inspire human composers or ultimately create independent AI musicians who can address perceptual and meaningful interactions with their human counterparts. Moreover, these smart creatures should be able to analyze and generate complex mathematical models that represent music, and recognize and understand musical sounds and human emotions (Roads, 1985). Although algorithmic music was already invented in the 1960s, it has only developed in the 21st century, as faster computing technology, deep learning, and artificial neural network have enabled a faster flow of information, and thus faster ways of problem-solving. This technology makes AI think more like a human, as something that can "learn by doing."

For example, researchers in Georgia Tech developed a robot named "Shimon" who cannot only compose but also improvise with many other human performers with its own composition on-the-fly. This AI Marimba percussionist is empowered by deep learning and a database of more than 5,000 songs to compose its own music. Moreover, a wide range of humanized body movements and gestures are also designed to empower Shimon, so it can make expressive bodily gestural communications with other human musicians during live performances (Hoffman and Gil, 2010). Zulić's article provides a detailed list of current AI musicians and companies as well as the history of this trend (2019).

It is also worth mentioning that AIVA,[30] created in 2016, was the first AI composer to be registered as an author in an author's rights society. Learning from 30,000+ scores in its database, AIVA has a plagiarism checker to prevent plagiarism from the training database and has passed many Turing tests with professional musician participants. People simply cannot tell if the songs are composed by AIVA or by humans. As a registered composer, with its own legal copyright and benefits, maybe AIVA can be seen as one of the answers to the historical debates around "cognitive labor" and the posthuman brain – where the brain labor does not necessarily have to be a human's – and we can see if society can treat this form of labor equally with other forms (Barrett, 2017).

Speaking of cognitive labor in the future of music-making, bio DMIs and data-driven DMIs should not be neglected. As early as in 1965, experimental composer Alvin Lucier premiered his piece *Music for Solo Performer for Enormously Amplified Brain Waves and Percussion* using electroencephalogram (EEG) electrodes to capture the performer's alpha brain waves' data and sonify them into various percussion sounds. These sounds were heard by the performer during the performance, which modulated the performer's brain activities in real-time, thus creating a biofeedback loop constructed through a network of neuroscientific cybernetics. Today, biofeedback DMIs can also be used in medical applications to save patients' life. For example, neuroscientists and music researchers at Stanford University have been working together to build an EEG sonification machine that makes obvious rhythmic periodic sonic patterns to monitor and detect the exact timing of a silent seizure-like activity. This way, the patient's further brain damage can be prevented (Parvizi et al., 2018). Extending this concept of data sonification, in this big data era data driven DMIs will be able to musically identify complex data patterns, thus helping extract the most useful information from massive datasets.

Last but not least, the current research on bio DMIs is not only investigating human and AI brainpower but also that of microbiological organisms. Miranda et al. (2011, 2016) have been using physarum polycephalum to realize sound synthesis, algorithmic compositions, and bio-computing. These little creatures have non-linear memory and can be used as memristors and processors in computing. As the NIME2019[31] annual conference keynote speaker, Miranda argued that "bio-tech is the new digital." He illustrated how *physarum polycephalum* serves as a voltage control mechanism, processes the information, and then gives feedback/response to the DMI system. Miranda calls this "biocomputer music." Again, biocomputer music and bio DMIs also involve debates about cognitive labor and the ethics behind this bio-art form and the bio-labor itself, as these tiny bio-memristors die very soon after they contribute their cognitive labor to us. Should we treat them simply as bio-electronic components, or simple organisms, or intelligent creatures that have lives and rights? Just like the discourse around robotic/AI labor, this phenomenon is also worth contemplating.

In many ways, posthuman DMIs lead to the change of human thoughts and behaviors in music-making. Although currently they cannot completely replace humans, they do help reduce the time that human musicians spend on repetitive compositional tasks (e.g., in gaming music); they provide inspiration and fast prototyping methods to help humans, especially laypersons, make music; they also create new genres and encourage new methods of composing. Where is this path ultimately leading? We don't know yet, as everything happens too fast to be fully understood. But one thing is certain: we must put humans (back) into this loop of creation.

Conclusion

Since the future needs to connect with the past and the present, multiple state-of-the-art current trends in DMI design and their brief histories have been introduced in this chapter. Space limitations prevent detailed discussions and references for all aspects. In other words, we see only the tip of the iceberg and the future may have millions of parallel directions which are unpredictable, in part because of many unforeseen factors.

However, musical dialogs between DMIs/NIMEs and acoustic music instruments will definitely be seen more often than not at formal concert settings in the very near future. Meanwhile, multimodal immersive DMIs that induce multisensory stimulations – encouraging human-centered design and learning, social connection, public health, and lasting well-being – will become the core principle of DMI design. On the other side, what we make, makes us. New human-computer interactions, biotechnology, big data, and embedded systems will create seamless connections between us and intelligent machines. New NIMEs, empowered by AI and other intelligent life forms, will merge with us as one inseparable posthuman.

On the sound generation side, with the rapid evolution of nature science and technology, new methods of synthesizing and transforming sound waves will shape the fundamentals of how we understand the sonic world. Because of the evolutionary digital audio technology, Curtis Roads was able to develop granular synthesis techniques and treat audio at a micro-level; understanding the nuts and bolts of physical properties of acoustic musical instruments led audio engineers to the success of developing physical modeling techniques to computer simulate these instruments in an authentic way that had never been previously achieved. As we learn more about the physical world, all the way down to a particle level, we will be inspired to come up with new ways to understand sound as a form of energy and matter. Currently, high energy physicists have already made significant progresses in understanding mysterious elementary particles such as the "God particle" – Higgs boson (Cho, 2012) and neutrino particles. One day, not too far away, such new scientific discoveries will provide critical analysis support and mathematical expressions of sonic phenomena both in the time and frequency domains, thus creating new sounds and transforming sounds from within.

Back in 1917, Edgard Varèse dreamed: "*Je rêve le instruments obéissant à la pensée* [I dream of instruments obedient to thought]." Only a hundred years later, we can almost make this dream come true. However, we are still on the path to the liberation of sound. This is because, in order to liberate sound, we must liberate our thoughts first, and then advance the technology that benefits our human species. Before we ask, "What can music technology do for us?" we should probably first ask ourselves, "What can our musicians do for us?" Before we ask, "How can we prevent superintelligence from wiping out human civilization?" we should probably first ask ourselves, "How can we create meaningful technology that advances humanity?" Such contemplation should be made to create a better future in our society, in arts, science, technology, engineering, mathematics, and humanities.

Notes

1 https://guthman.gatech.edu/
2 https://supercollider.github.io/
3 http://faust.grame.fr
4 https://puredata.info/
5 www.keyboardmag.com/artists/the-horizons-of-instrument-design-a-conversation-with-don-buchla
6 https://sensel.com/pages/buchla-thunder-overlay
7 https://buchla.com/easel-k/
8 www.modulargrid.net/
9 www.muffwiggler.com/forum/
10 https://github.com/VCVRack

11 www.makenoisemusic.com/
12 https://wmdevices.com/
13 http://4mscompany.com/
14 www.native-instruments.com/en/products/komplete/synths/reaktor-6/
15 https://madronalabs.com/
16 www.modulargrid.net/e/modules/evaluationlists
17 https://cycling74.com/
18 www.ableton.com/en/
19 https://ccrma.stanford.edu/software/jacktrip/
20 A free online tutorial, documentation, and course can be found at http://chrischafe.net/online-jamming-and-concert-technology-online-course/
21 https://plork.princeton.edu/
22 http://slork.stanford.edu/
23 www.smule.com/
24 https://new.steinberg.net/nuendo/virtual-reality/
25 www.dearvr.com/
26 www.youtube.com/watch?v=w76lNhqiaCw
27 https://donghaoren.org/blog/2018/buddhabrot
28 www.kuka.com/en-us
29 www.youtube.com/watch?v=WofWNcMHcl0&feature=youtu.be&t=391
30 www.aiva.ai/
31 www.ufrgs.br/nime2019/

Bibliography

Atherton, J., and Wang, G. *"Chunity: Integrated Audiovisual Programming in Unity."* NIME, 2018.

Barrett, G. D. "The brain at work: Cognitive labor and the posthuman brain in Alvin Lucier's music for solo performer." *Postmodern Culture 27*(3) (2017).

Bjørn, K. and Meyer, C. *Patch & Tweak*, 2nd edn. Bjooks, 2018.

Boulanger, R., and Lazzarini, V. *The Audio Programming Book*. MIT Press, 2010.

Braund, E., Sparrow, R., and Miranda, E. "Physarum-based memristors for computer music." In *Advances in Physarum Machines*, pp. 755–775. Springer, Cham, 2016.

Cabral, M., Montes, A., Roque, G., Belloc, O., Nagamura, M., Faria, R. R., Teubl, F., Kurashima, C., Lopes, R. and Zuffo, M. "Crosscale: A 3D virtual musical instrument interface." In *2015 IEEE Symposium on 3D User Interfaces (3DUI)*, pp. 199–200. IEEE, 2015.

Cáceres, J.-P., and Chafe, C. "JackTrip: Under the hood of an engine for network audio." *Journal of New Music Research 39*(3) (2010): 183–187.

Cecchetto, D. *Humanesis: Sound and Technological Posthumanism*, vol. 25. University of Minnesota Press, 2013.

Chadabe, J. *Electric Sound: The Past and Promise of Electronic Music*. Pearson College Division, 1997.

Cho, A. *"Higgs boson makes its debut after decades-long search."* Science (2012): 141–143.

Ciani, S. *"Making sounds with Suzanne Ciani, America's first female synth hero,"* interviewed by Kate Hutchinson. The Guardian, May 20, 2017.

Claude, C., and Wanderley, M. M. "Gesture-music." In *Trends in Gestural Control of Music*, M. M. Wanderley and M. Battier (eds). Ircam Centre, Pompidou, 2000.

Connor, N. O. *Reconnections: Electroacoustic Music & Modular Synthesis Revival*. Conference: Electroacoustic Music Association of Great Britain at University of Greenwich, London, 2019.

Cook, P. "2001: Principles for Designing Computer Music Controllers." In *A NIME Reader*, pp. 1–13. Springer, Cham, 2017.

Eduardo Reck, M., and Wanderley, M. M. *New Digital Musical Instruments: Control and Interaction beyond the Keyboard*, vol. 21. AR Editions, Inc., 2006.

Fillwalk, J. "Chromachord: A virtual musical instrument." *2015 IEEE Symposium on 3D User Interfaces (3DUI)*, pp. 201–202. IEEE, 2015.

Fukui, K., Ishikawa, Y., Shintaku, E., Honda M., and Takanishi, A. "Production of various vocal cord vibrations using a mechanical model for an anthropomorphic talking robot." *Advanced Robotics 26*(1-2) (2012): 105–120.

Gaudrain, E., and Orlarey, Y. *A FAUST tutorial*. Grame, 2003.

Glinsky, A. *Theremin: Ether music and espionage*. University of Illinois Press, 2000.

Good, I. J. "Speculations Concerning the First Ultraintelligent Machine." In *Advances in Computers*, vol. 6, Franz L. Alt and Morris Rubinoff (eds), pp. 31–88. Academic Press, 1965.

Hamilton, Robert, Smith, Jeffrey, and Wang, Ge. *"Social composition: Musical data systems for expressive mobile music."* Leonardo Music Journal (2011): 57–64.

Haridy, R. (2017). "Automatica – Robots that play drums, guitar and turntables and destroy a warehouse."

Available online at https://newatlas.com/nigel-stanford-automatica-music-robots/51258/ (last visited 6 March, 2018).

Hayles, N. K. *How We Became Posthuman: Virtual Bodies in Cybernetics, Literature, and Informatics.* University of Chicago Press. 2008.

Ho, E., de Campo, A., and Hoelzl, H. *"The SlowQin: An Interdisciplinary Approach to Reinventing the Guqin."* NIME 2019.

Hoffman, G., and Gil W. "Gesture-based human-robot jazz improvisation." *2010 IEEE International Conference on Robotics and Automation,* pp. 582–587. IEEE, 2010.

Jensenius, A. R., and Lyons, M. J. (eds). *A NIME Reader: Fifteen Years of New Interfaces for Musical Expression,* vol. 3. Springer, 2017.

Lahdeoja, O., Wanderley, M., and Malloch, J. Instrument augmentation using ancillary gestures for subtle sonic effects. In *Proceedings of the 6th Sound and Music Computing Conference, SMC '09,* pp. 327–330. Porto, Portugal, 2009.

Lui-Delange, K. W., Distler, S., and Paroli, R. "Sensel Morph: Product communication improvement initiative." 2018.

Martin, P. J. *Sounds and Society: Themes in the Sociology of Music.* Manchester University Press, 1995.

McCartney, J. "Rethinking the computer music language: SuperCollider." *Computer Music Journal, 26*(4) (2002): 61–68.

McLuhan, M. *"Understanding media: The extensions of man, 1964."* Pour comprendre les media: Ses prolongements. 1994.

Michon, R. and Smith, J. O. "Faust-STK: A set of linear and nonlinear physical models for the Faust programming language." In *Proceedings of the 14th International Conference on Digital Audio Effects (DAFx-11),* Paris, France, pp. 19–23. 2011.

Miranda, E. *Computer Sound Design: Synthesis Techniques and Programming.* Routledge, 2012.

Miranda, E. R., Adamatzky, A., and Jones, J. "Sounds synthesis with slime mould of Physarum polycephalum." *Journal of Bionic Engineering 8*(2) (2011): 107–113.

Parvizi, J., Gururangan, K., Razavi, B., and Chafe, C. "Detecting silent seizures by their sound." *Epilepsia 59*(4) (2018): 877–884.

Pinch, T. "Early synthesizer sounds." In *The Routledge Companion to Sounding Art,* p. 451. Routledge, 2016.

Pinch, T., and Frank T. *"The social construction of the early electronic music synthesizer."* Icon (1998): 9–31.

Puckette, M. S. *"Pure data."* ICMC, 1997.

Rich, R. *"Buchla thunder."* Electronic Musician. Aug 4, 1990.

Risset, J.-C. "The liberation of sound, art-science and the digital domain: contacts with Varèse, Edgard." *Contemporary Music Review 23*(2) (2004): 27–54.

Roads, C. "Research in music and artificial intelligence." *ACM Computing Surveys (CSUR) 17*(2) (1985): 163–190.

Roads, C. (ed.). *The Computer Music Tutorial,* 2nd edn. MIT Press, 1996.

Roads, C. *Microsound.* MIT Press, 2004.

Roads, C. *"History of electronic music synthesizers."* Unpublished manuscript. 2018.

Roads, C., and Mathews, M. "Interview with Max Mathews." *Computer Music Journal, 4*(4) (1980): 15–22.

Scaletti, C. "The Kyma/Platypus computer music workstation." *Computer Music Journal, 13*(2) (1989): 23–38.

Serafin, St., Erkut, C., Kojs, J., Nilsson, N. C., and Nordahl, R. "Virtual reality musical instruments: State of the art, design principles, and future directions." *Computer Music Journal 40*(3) (2016): 22–40.

Shanahan, M. *The Technological Singularity.* MIT Press, 2015.

Skeldon, K. D., Reid, L. M., McInally, V., Dougan, B., and Fulton, C. "Physics of the Theremin." *American Journal of Physics 66*(6) (1998): 945–955.

Smith, J. O. "Physical modeling using digital waveguides." *Computer Music Journal 16*(4) (1992): 74–91.

Sugano, S., and Kato, I. "WABOT-2: Autonomous robot with dexterous finger-arm – Finger-arm coordination control in keyboard performance." In *Proceedings, 1987 IEEE International Conference on Robotics and Automation, 4* (1987): 90–97.

Tanaka, A. "Mapping out instruments, affordances, and mobiles." In *Proceedings of the 2010 New Interfaces for Musical Expression,* pp. 88–93. Sydney, Australia, 2010.

Thorn, S. D. "Transference: A Hybrid Computational System for Improvised Violin Performance." In *Proceedings of the Thirteenth International Conference on Tangible, Embedded, and Embodied Interaction,* pp. 541–546. ACM, 2019.

Tung, B. T., and Schnieders, D. *"Pianow-Piano Self-learning Assistant in Mixed Reality."* 2018.

Vercoe, B. *"Csound."* The CSound Manual Version 3. 1986.

Wan, C. Y., and Schlaug, G. "Music making as a tool for promoting brain plasticity across the life span." *The Neuroscientist 16*(5) (2010): 566–577.

Wang, G. "Ocarina: Designing the iPhone's magic flute." *Computer Music Journal 38*(2) (2014): 8–21.

Wessel, D., and Wright, M. "Problems and prospects for intimate musical control of computers." *Computer Music Journal 26*(3) (2002): 11–22.

Wright, M. "Open Sound Control: An enabling technology for musical networking." *Organised Sound 10*(3) (2005): 193–200.

Wu, J. C. *"Experiencing Embodied Sonic Meditation through Body, Voice, and Multimedia Arts."* Ph.D. Diss., University of California Santa Barbara, 2018.

Wu, J. C., and Ren, D. *"'Resonance of the heart': A direct experience of embodied sonic meditation."* ISEA (2019): 247–254.

Wu, J. C., Yeh, Y. H., Michon, R., Weitzner, N., Abel, J. S., and Wright, M. M. *"Tibetan singing prayer wheel: A hybrid musical-spiritual instrument using gestural control."* NIME, 2015.

Xenakis, I. *Formalized Music: Thought and Mathematics in Composition.* No. 6. Pendragon Press, 1992.

Xiao, X., Locqueville, G., d'Alessandro, C., and Doval, B. *"T-Voks: The singing and speaking theremin."* NIME, 2019.

Zaripov, R. Kh. "An algorithmic description of a process of musical composition." In *Soviet Physics Doklady, 5* (1960): 479.

Zicarelli, D. "Max/MSP software." *San Francisco: Cycling '74.* 1997.

Zulić, H. "How AI can change/improve/influence music composition, performance and education: Three case studies." *INSAM Journal of Contemporary Music, Art and Technology 1*(2) (2019): 100–114.

Index

T - #0240 - 111024 - C0 - 254/178/13 - PB - 9780367362744 - Matt Lamination